573.2
L1995 Lampton
New theories on the origins
of the human race

DATE DUE 900775

Learning Resource Center
Marshalltown Community College
Marshalltown, Iowa 50158

NEW THEORIES ON THE ORIGINS OF THE HUMAN RACE

NEW THEORIES ON THE ORIGINS OF THE HUMAN RACE

CHRISTOPHER LAMPTON

900775

COMMUNITY COLLEGE LRC
MARSHALLTOWN, IOWA 50158

FRANKLIN WATTS
NEW YORK LONDON TORONTO SYDNEY
]1989[

Illustrations by: Vantage Art

Photographs courtesy of:
Photo Researchers: pp. 21 (Ulrike Welsch), 39 (Biophoto Associates Science Source), 65 (Tom McHugh), 69 (A. W. Ambler/National Audubon Society), 72 (Russ Kinne), 73 (Toni Angermayer), 83 (Jeanne White), 125 (D. W. Fawcett); The Granger Collection: pp. 23, 25, 46; Bettmann Archive: p. 37; American Museum of Natural History: pp. 34, 50, 51, 56, 59, 140; Magnum: pp. 97 and 99 (Ian Berry), 143 (Rene Burri); UPI Bettmann Newsphotos: p. 107; The Cleveland Museum of Natural History: p. 125; Institute of Human Origins: p. 118; Peter Jones: p. 134.

Library of Congress Cataloging-in-Publication Data

Lampton, Christopher.
New theories on the origins of the human race / by Christopher Lampton.
    p. cm.
Bibliography: p.
Includes index.
Summary: Presents recent theories on man's origins from both the study of fossils and the study of molecules.
ISBN 0-531-10783-3
1. Man—Origin—Juvenile literature. 2. Human evolution—Juvenile literature. [1. Man—Origin. 2. Evolution. 3. Man, Prehistoric.]
I. Title.
GN281.L35 1989
573.2—dc20   89-9169   CIP AC

Copyright © 1989 by Christopher Lampton
All rights reserved
Printed in the United States of America
5 4 3 2 1

CONTENTS

Introduction
] 9 [

PART ONE
BACKGROUND

Chapter One
The Tree of Life
] 15 [

Chapter Two
The Driving Force
] 32 [

Chapter Three
The Human Legacy
] 44 [

PART TWO
THE COMMON ANCESTOR

Chapter Four
The Missing Link
] 63 [

Chapter Five
The Molecular Link
] 78 [

PART THREE
THE FAMILY TREE

Chapter Six
The Leakey Bush
] 95 [

Chapter Seven
The Oldest Hominid?
] 113 [

PART FOUR
THE WAY WE ARE

Chapter Eight
Upright and Quite Bright
] 131 [

Notes and Sources Used 147
Recommended Reading 149
Index 153

NEW THEORIES ON
THE ORIGINS
OF THE HUMAN RACE

INTRODUCTION

Where did I come from?
   This may not be the very first question that a child asks, but it's one of the first. *Have I always been here? Where was I before I was here?*
   We all have a thirst to know the story of our origins, and not necessarily because we suspect that they may have something to do with sex. We want to know how we came to be on this planet, in this country, in this family. And even when we understand the process by which we came into existence as individuals, we still want to know where our parents came from, where their parents came from, and where our parents' parents' parents came from. The question *Where did I come from?* doesn't have a single answer; it has thousands, perhaps millions, of answers.

The biggest answer of all concerns itself not with our parents or grandparents but with the ancestors of all human beings everywhere on earth. It is the answer to questions such as *where did human beings come from, how did we end up on this planet,* and *what was here before we were?* The search for human ancestors is the ultimate quest for our roots.

Questions about the origin of the human race are the province of a special kind of scientist, one who combines the expertise of anthropologists—scientists who study human culture—and paleontologists—scientists who study the fossils of living organisms from the earth's prehistoric past. These scientists are known, appropriately enough, as *paleoanthropologists.*

As a science, paleoanthropology is barely more than a century old, and the answers that it provides, while fascinating, are still tentative. But in the last twenty years, it has been a science undergoing an exciting revolution. A new means of studying the human past has been discovered, one that goes beyond the study of fossils dug out of the earth. We can now examine the history of human evolution by reading a story written in the very molecules from which we are built. And the story that the molecules tell is not always the same as the one told by the fossils.

In this book, we'll talk about some of the answers to the question *where did we come from?* provided by paleoanthropologists, through both the study of fossils and the study of molecules. In the first three chapters, we'll discuss the history of the field. Then, in the rest of the book, we'll look at the ways in which scientists have answered crucial questions about how human beings became human.

PART ONE

BACKGROUND

ONE

THE TREE OF LIFE

Every person on earth is related to every other person on earth. In a real sense, we are all brothers, sisters, and cousins.

This is a simple truth, and one we often tend to forget. When we think of our relatives, we tend to think of our brothers and sisters—those people with whom we share common parents—and our first cousins—those people with whom we share common grandparents. We may even think about our third cousins—those people with whom we share common great-grandparents.

(The terminology of cousinhood is a bit odd. Second cousins are defined as the children of those with whom we share common grandparents—that is, the children of our first cousins—and thus we have skipped over them in the previous paragraph. But certainly they fit into the picture, too.)

After third cousins, though, the picture begins to get a little hazy. Surely we all have fifth and seventh cousins as well, those with whom we share common great-great-grandparents and great-great-great-grandparents, respectively, but we rarely know their names, and we are not at all likely to recognize their faces. Nonetheless, we might someday meet them at a family reunion and be surprised to learn that these people are our relatives, too. Perhaps some of them are people we are already familiar with from other contexts—they may be politicians or movie stars. Some of them may turn out to be people we will instantly like, or instantly dislike. But they are our relatives, nonetheless.

It is a common convention to think of families in terms of a *tree*, literally, a *family tree*. This doesn't mean that there is something about a family that is covered with brown bark and green leaves, but simply that there is a treelike quality to the way that a family divides into branches.

Figure 1, for instance, shows a small tree for a family that we will call the Oaktons. The current generation of the Oakton family consists of eight cousins named Mary, George, Susan, William, Robert, Louise, Ann, and Ronald. As is always the case with first cousins, they are related through a common pair of grandparents, who are represented by the uppermost branching point, or *node*, of the tree.

(*Node* is the technical term for both the ends and branching points of the lines in any treelike

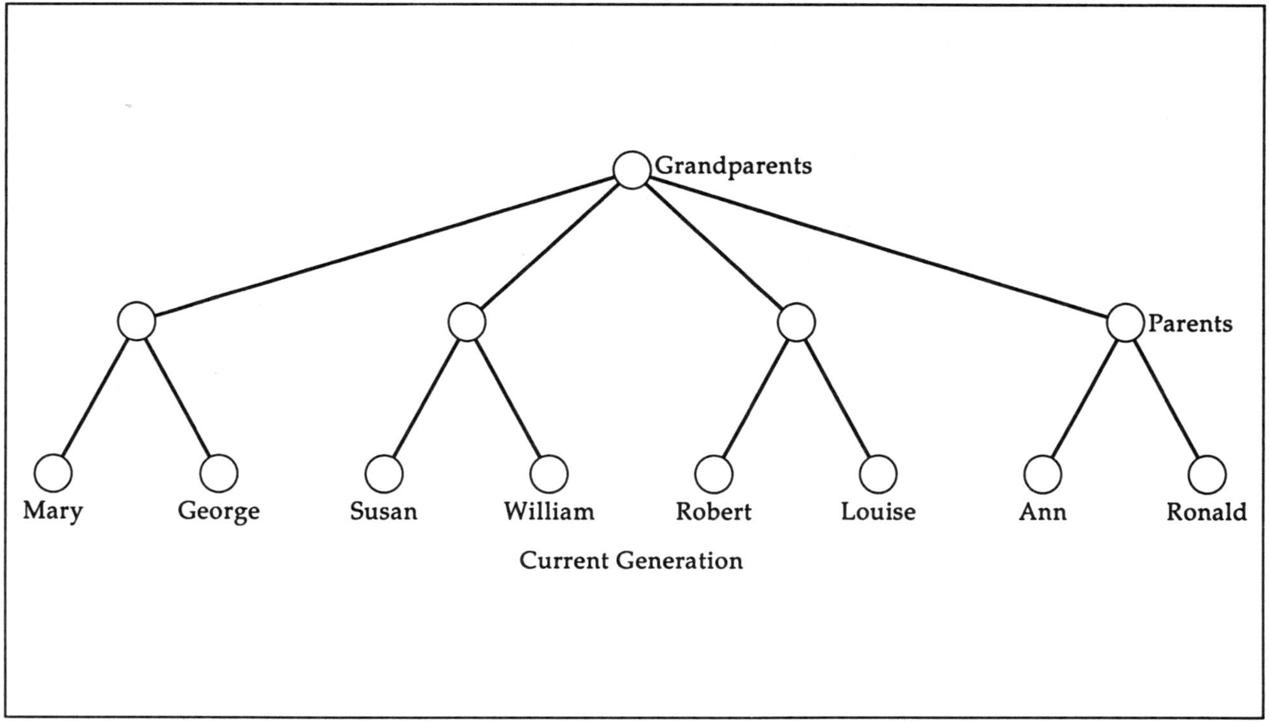

Figure 1

graph. We will be using the term frequently in this book.)

The parents of the eight Oakton cousins are represented by the nodes of the tree directly below the nodes representing the grandparents. The cousins themselves, the current generation of the Oakton family, are represented by the so-called *terminal nodes*—that is, the nodes at the very bottom of the tree.

Such a tree graph gives us a very clear idea of exactly how the various members of the Oakton family are related to one another. Perhaps you have seen similar tree graphs in history books, showing the relationships between the kings, queens, princes, and princesses of Europe, or the members of such American "royal" families as the Kennedys or the Rockefellers.

It's easy to see why such graphs are called "trees." Though they may lack bark and leaves, they have in common with trees the fact that the main branches split off into subbranches and sub-branches into sub-subbranches, and so forth. Furthermore, like real trees, such trees can grow, as further generations of the Oakton family are born, adding their own nodes to the tree.

We have shown you only a fraction of the complete family tree for our fictitious Oakton family. If we had included a node for great-grandparents above the node for grandparents, we could then include second and third cousins as well. A great-great-grandparent node would allow us to include fourth and fifth cousins. And so on.

Of course, we would eventually run into problems. We might, for instance, have to include in our tree people who do not have the last name Oakton but who nonetheless are direct blood relatives to the Oaktons. And, in fact, the naming conventions of Western society are such that the offspring of male members of the family bear the same surname, Oakton, though the offspring of female members are just as closely related.

This raises an important question. Where is the line between family and nonfamily? At what point do we say that these people aren't relatives of the Oaktons? Eighth cousins? Ninth cousins? Fourteenth cousins? Fifty-seventh cousins? Where do the Oaktons end and other families begin? At the point where the family name ceases to be Oakton?

If we allow the tree to contain blood relatives with names other than Oakton—and names, after all, are just an artificial convention, with no real basis in nature—then where is the dividing line between the Oakton family and the family of, say, the reader of this book? Where is the line between the reader's family and the family of the author of this book? Where is the dividing line between any two families?

Perhaps there is no such line. Perhaps in the widest, most extended sense, the Oaktons, the reader of this book, and the author of this book all belong to the same family: the family of humankind. If we enlarge our family tree sufficiently, we will discover that we all have a common relative, that we all have branched off from a single node of a single family tree.

In this sense, the whole world is our family, since we share a common relative with everyone who was ever born, from Julius Caesar to the Queen of England to the star of the latest blockbuster motion picture. Go back far enough, and you'll find that you have the same great-great-multi-great-grandparents as any other person you will ever meet or hear about. In that sense, everybody is your cousin, though some are more distant cousins than others.

This may come as a surprise. Cousins tend to look alike, yet people (we think) often look quite different. Some have pale pink skin, others dark brown skin, still others skin with a hint of yellow or a ruddy redness. Some people have curly black or red hair, others straight blond or brown hair. Some have broad noses, others narrow noses. And so on and so on. How can all these people be cousins? They don't even look alike!

But the truth is, for all the variations in physical appearance found throughout the human race, we are all of us much more alike than we are different. We all have the essential qualities of humanness—in particular, our large brains and our erect postures—that set us apart from all other animals on the earth. We all feel the emotions of love and anger, curiosity and fear, pride and joy. We all have brains that can add two numbers, read the words in a book, and recognize the face of a relative. In a real way, the human race is one big family.

Does the extended family of humanity end there? We are all cousins, yes, but what of the other living things on earth? Do they belong to another family? Are they completely separate from the human family?

At first blush, this would seem to be the case. When we look at other human beings, at least we recognize the family resemblance. We see faces that have the same essential features as ours, arranged in the same essential way. We see bodies that can achieve the same postures that ours can, hands that can grasp in the same way ours can, mouths that can chew the same food as we do.

When we look at species other than our own, all traces of resemblance seem to vanish. In many instances, these other organisms do not even have faces, much less faces that bear a resemblance to ours. And even those animals that do bear a resemblance to human beings—apes and monkeys, for instance—differ from us in bizarre and startling ways. They cannot walk erect, except in short, stumbling bursts. They are covered with thick fur, and they cannot speak, except in incomprehensible bursts of noise. Yet these are the creatures who resemble us most of all! Surely a creature such as a grasshopper or an amoeba resembles us not in the slightest!

But that isn't so. Even the "lowliest" of creatures—and the term "lowly" in this context is an unfair human judgment that in no way does justice to the sophisticated construction of even the simplest bacterium—resembles us in more ways than we might imagine. Surely, we are more like a

*What of the other living things on earth?
Are they separate from the human family?*

bacterium, say, than the bacterium is like a rock, a stream, or any nonliving thing.

What do we have in common with a bacterium? A surprising amount, actually. The human body (and the bodies of every other animal large enough to see with the naked eye) is made up of tiny structures of living material called *cells*. And each of these cells bears a striking resemblance to a bacterium.

The bacterium is made up of the same types of *molecules*—large groupings of atoms—that the human cell is. It reproduces itself in much the same way, by splitting in two and bequeathing duplicate copies of itself to both offspring. It contains much of the same "machinery," and that machinery operates in much the same way. Thus, at the most basic level, we bear much more resemblance to a bacterium than you might think.

The idea that there is a kind of family resemblance between all of the living things on earth is not a new one. People have been noticing these resemblances for thousands of years. But it was only two hundred years ago that the study of these resemblances became a science. The science was—and is—called *taxonomy*. Its founder was the great Swedish scientist Carolus Linnaeus.

Linnaeus believed that we could use the obvious resemblances among living organisms as a means of classifying those organisms, the way that merchandise in a store is classified by the way in which it is used or books in a library are classified by subject matter.

The Swedish physician and botanist Carolus Linnaeus (1707–1778) developed a system of classifying organisms.

This idea did not originate with Linnaeus. A system for the classification of living organisms had existed for at least a century before Linnaeus entered the picture, but it was clumsy and inconsistent at best. It involved giving all organisms a so-called *polynomial name,* consisting of a string of Latin phrases perhaps fifteen or sixteen words in length. The first word in the polynomial name was the *generic name,* that is, the name that identified the *genus* (group) to which the organism belonged, such as *Felis* (cats), *Equus* (horses), *Quercus* (oak trees), and so forth. The remaining words in the polynomial name constituted a fairly detailed Latin description of the organism and served to differentiate it from all other members of that genus.

For instance, there was a certain kind of oak tree for which the polynomial name was *Quercus foliis lanceolatis integerrimus glabris.* The first word identifies the tree as a member of the oak genus *Quercus.* The rest of the name says that this particular oak has "spear-shaped, smooth leaves with no teeth along the edges." This differentiates that particular type of oak from other types, for instance, *Quercus foliis obtusinuatis setaceo-mucronatis,* or "oak with leaves with deep blunt lobes bearing hairlike bristles." Quite a mouthful!

Obviously, these polynomial names were hard to remember, awkward to use, and frequently difficult to pronounce! Linnaeus quietly replaced this system of *polynomial nomenclature* (nomenclature is just another word for "name") with a new system of *binomial nomenclature,* which caught on immediately. In Linnaeus's new system, every species of organism was identified by only two names, the generic name (which was usually the same one used in the polynomial system) and the *species name,* which belonged to that species and no other. (Occasionally a third name, representing a *subspecies,* is added to the binomial name to differentiate closely related but subtly different organisms.)

Thus, the first kind of oak mentioned above became *Quercus phellos* and the second kind became *Quercus rubra.* (Note that the first letter of the first word of a generic name is always capitalized; the species name is never capitalized, and scientific names are italicized because they are in Latin.) No longer did botanists or anyone else who studied living species have to remember tongue-twisting polynomial names. After Linnaeus, two names would do. And we still use Linnaeus's system of binomial nomenclature today.

Linnaeus also popularized the idea that organisms could be grouped into higher ranks than genera (pl. of *genus*). Today we have an extremely detailed method of classifying living organisms into groups and supergroups, thanks in large part to the enduring influence of Linnaeus.

For instance, we now group related genera together into larger groups called *families.* Families are, in turn, grouped into *orders,* orders are grouped into *classes,* and classes are grouped into *phyla* (pl. of *phylum*). Finally, phyla are grouped into the largest units of all: *kingdoms*. Those organisms large enough to be seen with the naked eye are grouped into two huge kingdoms, the *Plantae* (plants) and the *Animalae* (animals). Plants are those organisms that obtain their energy directly from the sun through the process of photosynthesis. Animals are those organisms that must obtain their energy by eating other organisms, such as plants and other animals.

To give but one example, the common honeybee is known in the binomial nomenclature as *Apis mellifera*, which tells us that it is a member of the genus *Apis* and the species *mellifera*. But it is also a member of the family *Apidae*, the order *Hymenoptera*, the class *Insecta*, the phylum *Arthropoda*, and the kingdom *Animalia*.

The classification of an organism in this system is based on its physical resemblances to other members of a particular group. Put another way, all animals grouped together have one or more physical features in common. For instance, members of the phylum *Chordata* all have a notochord (precursor of the backbone) at some time in their development. Members of the class *Mammalia* all have mammary glands, hair, and certain other characteristics. And so forth.

Like all other animals, human beings have a place in this system. In binomial nomenclature, we are called *Homo sapiens*, meaning "man the wise." We belong to the family *Hominidae*, the order *Primates*, the class *Mammalia*, the phylum *Chordata*, and the kingdom *Animalia*.

Technically, this type of system is called a *hierarchical system*. Interestingly, such a hierarchical system can also be represented by a tree-shaped graph, just as a human family can, except that in the graph of a hierarchical system, the nodes represent categories such as orders and phyla instead of parents and grandparents. Figure 2 shows an extremely simplified tree graph representing the relationship between several animals.

Is it just a coincidence that this graph bears such a striking resemblance to a family tree? Or is it possible that somehow all of the organisms represented by this hierarchical system are in some

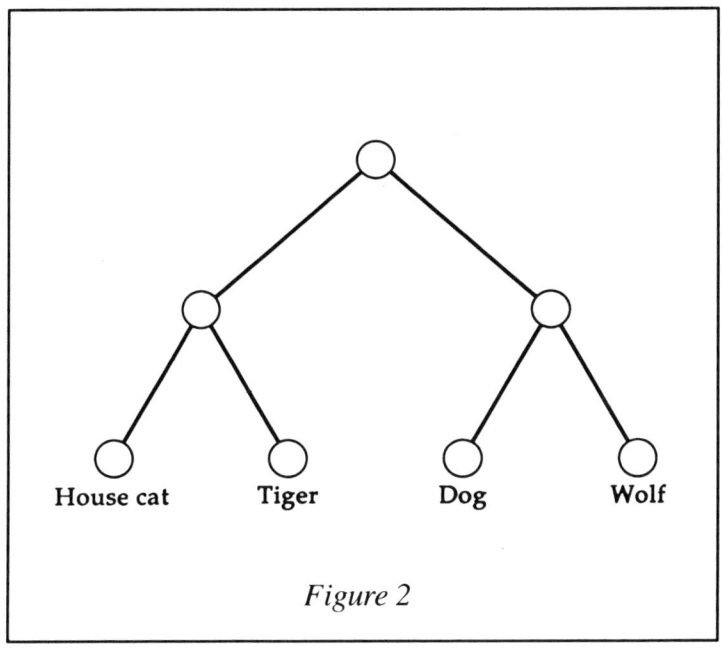

Figure 2

sense blood relatives, that they are related much as members of a human family are related? Is it possible, in short, that the nodes on the hierarchical tree of life represent common ancestors of various species the way that the nodes on a family tree represent common grandparents? Could the various species within a classification have inherited their common features—spinal cords, mammary glands, etc.—from these common ancestors? And, if so, where *are* the common ancestors that the nodes represent? Do they still exist? Or did they become extinct at some time in the past?

Like most people of his time, Linnaeus believed that God had created each species of organism separately, during the six days of Creation. Thus, there could be no blood relationship, and therefore no common ancestors, between the various species that he and others were classifying. Further, Linnaeus held the common belief that the Creation had taken place in the year 4004 B.C. Even if it were somehow possible that blood relationships existed between species, there had been insufficient time for the common ancestors to become extinct. And since no common ancestors between living species were found in the modern world, they could not possibly exist.

But by the end of the eighteenth century, it was becoming increasingly obvious that there was an error in this way of thinking. Geologists—scientists who study the very rocks of the earth—had begun to theorize that the earth was much older than a mere 6,000 years; that, in fact, the age of the earth might be measurable in the millions or even billions of years. And paleontologists, scientists who study the preserved remains of organisms that lived on earth long ago, were uncovering evidence that the organisms that lived on the earth thousands and millions of years ago were very different from the organisms of today.

It was possible, therefore, that the nodes of the tree of life actually *did* represent common ancestors of living creatures, ancestors that lived on earth long ago but had subsequently ceased to exist. Yet that left an even more perplexing riddle to answer: How is it possible for one type of organism to give birth, in effect, to a different type of organism? Human cousins tend to look pretty much like their grandparents, and pretty much like each other, for that matter. At the very least, they all tend to have two eyes, a nose, a mouth, two arms, two legs, and so forth. But the same thing can hardly be said of a bacterium and a human being, to cite but two out of several million possible examples. Yet the tree of life suggests that they have a common ancestor. How can such a thing be?

The answer came in the nineteenth and again in the twentieth century. It was supplied in the nineteenth century by Charles Darwin, arguably the greatest biologist of all time, and in the twentieth century by scientists known as geneticists and molecular biologists. It was an answer that told us something startling and quite unexpected about the human race itself.

TWO

THE DRIVING FORCE

What makes the tree of life grow? We know that living organisms reproduce—that is, produce other living organisms—but how can those living organisms change with time? If we are to believe that the tree of life is a genuine family tree, that all living things have common relatives, then how did the different species of living organisms come to look so different? Why isn't everything alike?

In the eighteenth and nineteenth centuries, this question was being asked by many educated people. An exciting idea was in the air, an idea called *evolution*. The idea was a simple one: that given sufficient time, living species tended to evolve—change slowly—into other species. But why? What is the driving force behind this slow process of change?

This question was answered in 1859, in a book

entitled *On the Origin of Species by Natural Selection, or the Preservation of Favoured Races in the Struggle for Life*. Its author was Charles Darwin.

As the title of the book implies, Darwin suggested that the driving force behind evolution was a process called *natural selection*. All individuals in a species differ somewhat from one another—such as in height, body build, hair color, eye color, etc. These natural *variations* result from sexual reproduction, when the genes of both parents are "mixed" to produce the offspring. Darwin was aware of these variations. Also, like many biologists before him, Darwin had noted that occasionally an organism would be born with features that were in some way different from those of the parent organisms and in some cases even different from the features of all other members of the species. These new features, called *mutations,* occur randomly in a species.

Could variation be the driving force behind evolution? No; its very randomness prevents us from seriously considering mutations as the cause of evolution. The chances that organisms as complex as human beings (or even as complex as bacteria) could evolve as a result of random variations are so slim that such an eventuality can be considered effectively impossible.

But Darwin noticed that there were other forces acting in the environment in which an organism lives that can select those variations that increase the organism's chance of surviving within that environment. These forces, combined with

Charles Darwin (1809–1882) developed a theory of evolution with the aim that "light will be thrown on the origin of man and his history."

variation, could serve to explain evolutionary change.

Simply put, Darwin reasoned that if an organism inherits a random variation that decreases its chances of surviving within the environment in which it lives, then the organism will have little chance of reproducing and passing the mutation to its offspring; thus the variation will be "selected" out of the species. But if, by sheer chance, a particular variation improves the organism's chances of surviving within its natural environment, then the organism will probably reproduce and not only pass the variation onto its offspring but will probably also live long enough to have more offspring than other members of the species who do not possess the variation. Eventually, the offspring of that organism will become more numerous in the species, and the change will be spread far and wide. The species will have changed, in some cases actually becoming a brand-new species, through accumulation of small changes over long periods of time.

This combination of natural variation and environmental "selection" was given the name natural selection by Darwin. Unlike variation itself, natural selection is not a random process and is quite capable of explaining evolution. Again and again, biologists have shown that natural selection is sufficient to cause the evolutionary changes that would explain how species as diverse as human beings and bacteria could have descended from common ancestors, given that millions and even billions of years have passed since those common ancestors existed.

Although natural selection explained the process of evolution, it still left some unanswered questions. Why did mutations occur in the first place? Why were they passed on from generation to generation?

This question was being answered even as Darwin published his theory, but the answer was ignored for about forty years, until the beginning of the twentieth century. In the 1860s, an Austrian monk named Gregor Mendel developed a new science that later came to be known as *genetics.* He proposed that every living thing possesses a number of factors—we now call them *genes*—that determine all of its physical characteristics. A given human being, for instance, possesses not only genes determining color of hair, color of eyes, length of nose, etc., but also genes for all the things that make that person human, genes for standing erect, for instance, and genes for a large brain—even genes for having skin, lungs, bones, and so forth. These genes are passed down from parent to child, and thus children tend to have the same eye color, hair color, etc., as their parents, or at least their grandparents. (The physical characteristics determined by genes can sometimes skip a generation or two, according to rules that will not be considered here.)

Mendel's theory of genetics was published in an obscure magazine and ignored for forty years. It was independently rediscovered by several re-

Gregor Mendel (1822–1884), Austrian botanist and Augustinian monk, began the science of genetics in experiments with peas in the monastery garden.

searchers around the year 1900 and eventually became part of a revised version of Darwin's theory, called the *Modern Synthesis,* in the 1930s. The Modern Synthesis says that the changes in a species that are selected by natural selection are the result of spontaneous changes in the genes.

But what are genes? To Mendel they were mysterious factors that were passed down from generation to generation, but surely modern science must have more to say about this subject than Mendel did.

In fact, modern science has a great deal to say about this subject, and what modern science has to say about this subject has had a profound impact on the study of human evolution, as we shall see in later chapters of this book.

According to modern biologists, the genes are actually contained in the individual cells of the bodies of living things. These cell structures, called *chromosomes,* are made up of molecules—chains of atoms—of a special substance called *deoxyribonucleic acid, DNA* for short. Each cell of your body contains forty-six chromosomes. Other organisms have similar collections of chromosomes within their cells, though the specific number of chromosomes may vary with the species.

Each chromosome contains the instructions for assembling and operating the body of the organism of which it is part. How can chromosomes contain information? In effect, the information is "written" on the DNA molecule in the chromosome in the form of sequences of smaller molecules

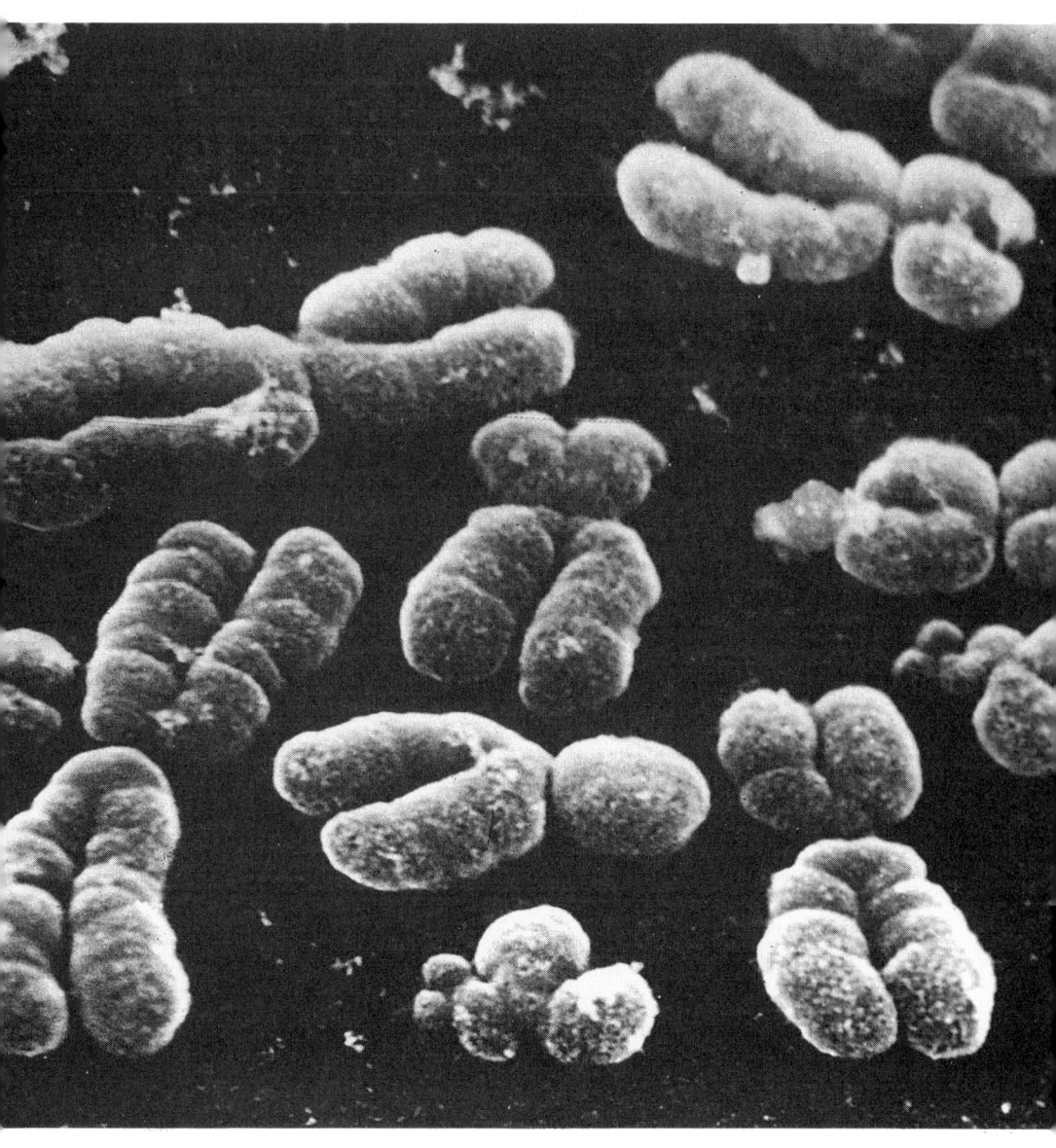

Human chromosomes contain instructions for assembling and operating the body of the human organism.

that spell out the information the way that the words and letters of the sentence you are now reading spell out the information in this paragraph.

This molecular alphabet, however, is a great deal simpler than the twenty-six-letter alphabet in which this book is written. In fact, the molecular alphabet has only four letters: A, G, C, and T. (Actually, these letters correspond to the small molecules of the substances known as *adenine, guanine, cytosine,* and *thymine,* though you need not memorize these names.) The cells of living organisms contain special equipment for "reading" the information in these molecules and translating it into the necessary actions, the way that a computer contains equipment for translating a computer program into an information-processing action. Specifically, the machinery converts the genetic information into protein molecules, tiny molecular machines that perform all the tasks involved in building and running a living organism.

You need not worry about the way in which the genetic information is spelled out by these "letters." If you are curious, there should be several good books on genetics and molecular biology at your local library. (See the Recommended Reading list at the end of this book for suggestions.) Just keep in mind that it is these instructions, contained in the DNA molecules, that determine just what species an organism belongs to. In a very real way, the only difference between you and a bacterium is information, the information in your genes and the information in the genes of the bacterium.

When an organism reproduces, the information in its DNA is carefully copied and passed on to the new organism, which is why organisms are always the same species as their parent or parents. (In the case of a bacterium, which has only a single parent, this information is an exact copy of the information in its parent's genes. In the case of human beings and other organisms that have two parents, this information is a mixture of information from both the mother and father.)

Occasionally, something goes wrong in this copying process, and some of the genetic information passed on to the offspring is incorrect. For instance, suppose that the parent has a gene that reads like this:

ACCGTGAAGTGACGGT

We might suppose, for the sake of argument, that this is a highly simplified version of one of the genes that makes eyes blue (most characteristics are influenced by more than one gene). When the information in this gene is passed on to the offspring, however, it might be slightly miscopied, like this:

ACCGTGTAGTGACGGT

If you look closely, you'll see that one A has become a T. Is this enough of a change to turn one species into another species? Hardly. In fact, it may not be enough of a change to make any difference at all, the way that a single typographical error in a sentence is rarely enough to change the meaning of the sendence, um, sentence.

But it's possible that an error such as this, occurring in a gene that, say, codes for blue eyes, will be sufficient to give the bearer of the gene a color of eyes that has never existed in the species before, perhaps chartreuse. And if it turns out that chartreuse eyes give that individual an advantage within its environment—by helping it to see better in dim light, for instance—then the gene may be passed on to an unusually large number of descendants and thus become a dominant feature of the species. On the other hand, if the gene puts its bearer at a disadvantage within its environment—by interfering with its ability to see in dim light, for instance—then the gene may not be passed along at all.

It takes a long time for one species to change into another species by natural selection. But if there is one thing that living organisms have in abundance, it is time. According to current theories, life originated on this planet almost four billion years ago, and it has been evolving through natural selection ever since. Early living organisms probably resembled a naked DNA molecule floating in the primeval ocean. More advanced organisms would have resembled modern bacteria. Eventually, multicelled organisms evolved, first as colonies of independent cells and then as collections of cells that could no longer survive without each other's mutual assistance. Human beings are just such a collection of cells, as are cats, honeybees, oak trees, and all other organisms large enough to be seen with the naked eye.

Thus, the great tree of life really is a family tree. Just as the current generation of the Oakton family represents the terminal nodes on the Oakton family tree, so the living organisms that populate the earth today are the terminal nodes on the evolutionary tree. Every living thing on earth is your cousin, in much the same sense that every other human being is your cousin. The relationships are more distant—it has been at least a billion years, for instance, since the common ancestor of humans and bacteria was alive—but they are relationships, nonetheless.

However, just as we feel a special kinship for those in our immediate family that we do not feel for the people who pass us each day on the street, it is natural that we should be especially interested in the members of our own species—our evolutionary family—and how it came into existence. Thus, we will now move from the subject of general evolution to the more specific topic of human evolution. We will move from the tree of life to the specific branch on that tree that led to the existence of our own species, *Homo sapiens.*

THREE

THE HUMAN LEGACY

It took Darwin nearly thirty years to write *On the Origin of Species*. Much of this time was spent researching the evidence for natural selection and building solid arguments in favor of the theory. But time and again, Darwin postponed work on the *Origin* in favor of less important books. And he might never have published it at all if he hadn't received word that another scientist had independently arrived at the theory of natural selection and was about to publish a book about it himself.

Why did Darwin repeatedly postpone publication of the *Origin*? If his theory was so important—and certainly Darwin was acutely aware of its importance—why didn't he rush to put it in print as soon as possible?

We can only guess, but it would seem that Darwin was actually afraid to publish the *Origin* because he knew that it was going to be controversial. And it was going to be controversial because it led to the inevitable conclusion that human beings had descended from a common ancestor with the apes.

It was bad enough that the theory of evolution contradicted the biblical account of Creation. (Darwin, incidentally, believed that living organisms were created by God but that natural selection was the tool that God had used and that it had taken much longer than the six days reported in the Bible.) But as long as evolution was not totally understood, it was possible to believe that even if evolution itself was a fact, human beings might be exceptions to it. Perhaps other organisms evolved, but human beings had been specially created.

By detailing the mechanism of evolution, Darwin's theory removed the last ounce of doubt. Everything was subject to natural selection, including human beings. The apes were our closest relatives on the tree of life, and thus they were the ones with whom we had most recently shared a common ancestor. They were, in short, our evolutionary first cousins. Darwin knew that although this idea might be acceptable to most scientists, it was not going to sit well with the general public.

He was right. Even though he hardly mentioned human beings in the *Origin*, except to say rather coyly that thanks to his theory, "light will be thrown on the origin of man and his history," anyone who looked at the tree of life fashioned by Linnaeus and others knew the implications of Darwin's theory.

Many newspapers and magazines reflected the public's dismay at Darwin's theories. In this 1871 cartoon by Thomas Nast, a gorilla seeks the protection of Henry Bergh, founder of the American Society for the Prevention of Cruelty to Animals.

The human position on the tree of life is an interesting one. Our binomial name is *Homo sapiens*, which tells us immediately that our species is a member of the genus *Homo*. In fact, we are the only living member of this genus. Unlike, say, housecats (*Felis domesticus*), who share the genus *Felis* with lions (*Felis leo*), tigers (*Felis tigris*), panthers (*Felis panthera*), and so forth, or dogs (*Canis familiaris*), who share the genus *Canis* with wolves (*Canis lupus*), we share the genus *Homo* with no other species. We are all alone on the *Homo* branch of the tree of life, the sole terminal node branching off from the *Homo* node.

Furthermore, we also have our own family, *Hominidae*, the next node on the tree of life above the genus *Homo*. Once again, we are the sole members of this family, the only terminal node branching off from the *Hominidae* node. In this regard, we are unique. Linnaeus himself wondered, late in life, if this classification had not been a mistake. The resemblance between humans and apes was so great that he felt he perhaps should have classed humans and apes in a single family.

With *Homo sapiens* being the only species that branches off from both the *Homo* and *Hominidae* nodes of the tree, it seems strange to refer to these as nodes at all. A node is either the end of a branch or the point at which a branch splits into two or more new branches. But these nodes are neither. Why bother to have these nodes on the tree at all?

That's a difficult question to answer, since the decision to assign classifications such as genera and families can sometimes be a trifle arbitrary, as Linnaeus's doubts about his own classification of *Homo sapiens* indicate. Nonetheless, is it possible that although these nodes may no longer be branching points on the tree of life, they once were? In other words, could these nodes represent branches on the human family tree that no longer exist, species that long ago became extinct?

We now know, of course, that this is precisely what they do represent, but in the time of Linnaeus and Darwin there was no compelling evidence for this. What might have helped to settle the controversy over *On the Origin of Species* was some kind of tangible evidence of the human and prehuman past, specifically the fossil remains of prehistoric human beings, individuals who had existed at an earlier point in evolutionary history. And the most sought-after fossil in Darwin's day was the one known popularly as the "missing link," the common ancestor of humans and apes.

The idea of fossils—the preserved remnants, usually bones, or imprints of organisms that lived long ago—had been around before Darwin's time. But the idea that the earth was at least hundreds of millions of years old was only just beginning to be accepted, and the idea that the earth itself held the bones of species that no longer lived on the planet was still controversial. Yet if humans had descended from a common ancestor with apes, then somewhere in the so-called fossil record (the layers of fossils buried in the earth, the oldest in the deepest layers, the most recent in the uppermost layers) should be the tangible proof. These fossils had the potential to resolve the controversy over human evolution. Not surprisingly, many fossils discovered since (and even before) Darwin published his theory have themselves been the object of much controversy, even among the scientists who study them.

In fact, the first fossils of prehistoric human ancestors to be later identified as such had been discovered even before Darwin had published his theory. A prehistoric human skull had been unearthed from the Rock of Gibraltar in 1848, and, half a century earlier, rocks that appeared to be crude stone tools had been found along the Somme River in France, though the suggestion that they might have been carved by ancient human beings was widely regarded as absurd.

One of the most dramatic of such discoveries was made in 1856—only three years before Darwin's book was published—in the Neander Valley of Germany. It was the partial skeleton of a man, or something very like a man, found buried in a cave. The skull was abnormally thick, and the ridges of the eyebrows were unusually protruding, but almost certainly it was human.

However, it was not the missing link. It was too humanlike to be the common ancestor of apes and humans. And yet some scientists thought that it was not quite a modern human being, either. Because it had been found in the Neander Valley, some scientists began referring to it as *Neanderthal man* and suggested that it might be a prehistoric ancestor of modern humans.

This idea was attacked viciously by several other scientists of the period. One declared in all

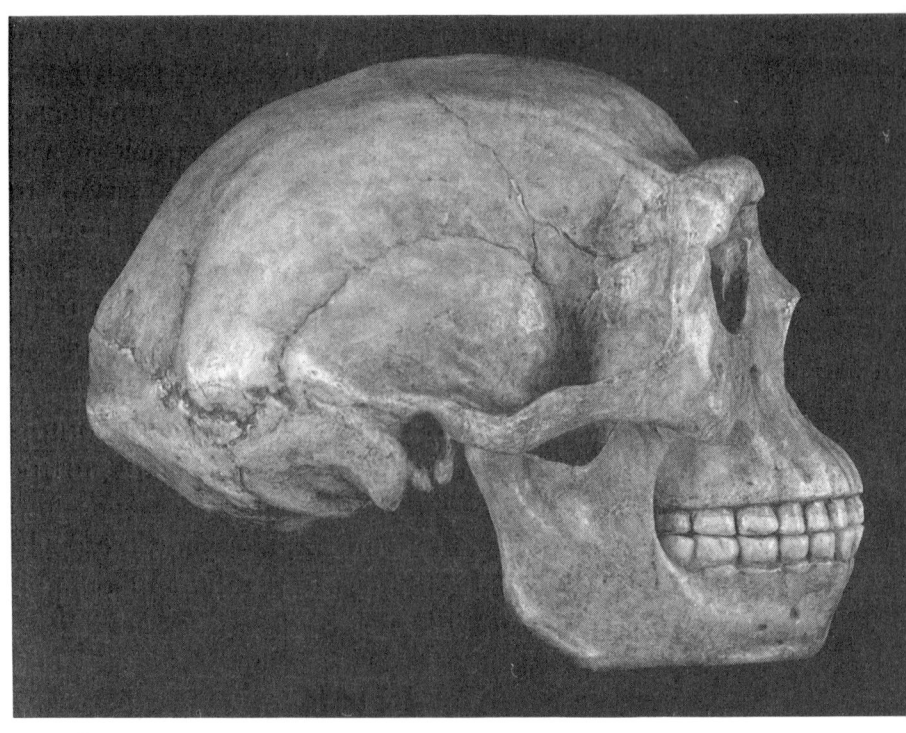

A reconstructed Neanderthal skull (above) and (opposite) an early restoration showing what a Neanderthal may have looked like.

seriousness that the skeleton was that of a Cossack who had fought in the Russian army during the Napoleonic Wars. Separated from the other troops, perhaps sick and wounded, he had retreated to the cave for shelter and died. Another scientist declared that Neanderthal man was a modern human being who had suffered from a deforming disease.

Eventually, Neanderthal man would come to be regarded as a slightly earlier form of human, though not necessarily a direct ancestor of modern human beings. The species was given the scientific name *Homo sapiens neanderthalensis,* which places it in the same genus and species as modern *Homo sapiens*—that is, the genus *Homo* and the species *sapiens*—but in its own subspecies, *neanderthalensis.* To distinguish us from *neanderthalensis,* we were also given our own subspecies, *Homo sapiens sapiens.* (Roughly translated from the Latin, this means "doubly wise man.") *Neanderthalensis* is now known to have lived in Europe from about 100,000 to 35,000 years ago.

The next major find in the search for fossil human ancestors came about a decade later, in the Cro-Magnon region of France. For this reason, the fossils found were dubbed *Cro-Magnon man,* though they are now regarded as fully modern human beings, prehistoric members of our own subspecies *Homo sapiens sapiens.* Cro-Magnon seems to have first appeared in Europe about 35,000 or so years ago.

No more major finds were made until the last decade of the nineteenth century. Eugene Dubois, a medical doctor from Holland, was fascinated with the idea of discovering a missing link. He reasoned that if human beings had descended from apes, then the remains of prehistoric human beings would be found not in places such as France or Germany but in the places where apes are found today, in the tropics. He joined the Dutch East India army and arranged to be stationed in Sumatra, in the East Indies. There he searched for the remains of prehistoric humans.

Perhaps as much by luck as through astute reasoning (Darwin, for instance, had believed that humans descended from the apes of Africa, not the apes of the East Indies), Dubois found what he was looking for. Reposted to Java after an illness, he persuaded the army to give him a detachment of convict laborers to dig for fossils on the banks of the Solo River.

At first, the laborers sold the fossils to black-market dealers as rapidly as they could dig them up (it was a common belief in China that fossils were the bones of dragons and had magical curative powers), but Dubois soon solved this problem and amassed a huge collection of fossils, a few of which seemed to be human.

The most promising of these were a skullcap, a thighbone, and some teeth. Dubois believed that they were the remains of a prehistoric apeman—a missing link between humans and apes—that he called *Pithecanthropus erectus,* which is Latin for "erect-walking apeman." Today, paleoanthropologists believe that *erectus* is, in fact, not an "apeman" but a member of the same genus, *Homo,* as modern human beings; thus, the species has been renamed *Homo erectus*—"erect-walking man." For many years, the fossil was popularly known as *Java man.*

It was quite a long time, however, before Dubois's fossil was accepted by other scientists. It remained controversial for many years. Embittered by the controversy, Dubois eventually buried the fossils under the floor of his dining room and refused to let anyone see them.

In 1912, an amateur geologist in England named Charles Dawson found the skull and jaw of what appeared to be a prehistoric human in Piltdown, Sussex. Popularly known as the *Piltdown man*, the fossil was named *Eoanthropus dawsoni* ("Dawson's apeman") by scientists.

Piltdown man confirmed a prejudice held by some scientists of the period that early human beings possessed large brains and apelike jaws. It also confirmed the prejudice of many English paleontologists that early human beings had lived in prehistoric England.

Alas, Piltdown man turned out to be a hoax, not a genuine fossil at all. Some unknown prankster had combined the skull of a modern human being with the jaw of an ape, stained them to make them look old, and buried them where Dawson would find them. The hoax was not revealed until the early 1950s, more than forty years after the original discovery of the fossil. The perpetrator of the hoax has never been identified, though at one time or another nearly everyone involved with the fossil, including Dawson himself, has been accused of the deed.

One of the most important paleoanthropological discoveries—the discovery of an entirely new *genus* of human ancestors—was made in 1924, in Johannesburg, South Africa, but it remained controversial for another two decades. The discovery was not made in a dark cave or fossil dig; it was made in a well-lighted room by an anatomist searching through a box of fossils that had been sent to him by a friend.

The anatomist was Raymond Dart. The fossils had come from a lime quarry in Taung, in Bechuanaland (now South Africa). Dart had received several such boxes of fossils from a friend, and, while sorting through one of the boxes, he came across a skull that was interesting, indeed.

(Incidentally, the afternoon on which Dart received these boxes of fossils was the same afternoon that he was to be best man at another friend's wedding. Dart, half dressed in his tuxedo, became so absorbed in the fossils that he forgot all about the wedding. At the last minute, his friend arrived in a panic and practically had to drag Dart away from the bones and force him to finish dressing.)

The fossil that had caught Dart's attention appeared at first glance to be that of a baby ape. Dart, however, noticed a couple of things that were unusual about it. For one thing, the braincase—the part of the skull that once contained the brain—seemed rather large and unusually shaped for an ape. For another, the *foramen magnum*—the opening through which the spinal cord enters the skull—was in the wrong place for an ape. In an ape, the spinal cord enters at the rear of the skull, as befits the "stooped" posture of the ape. (The

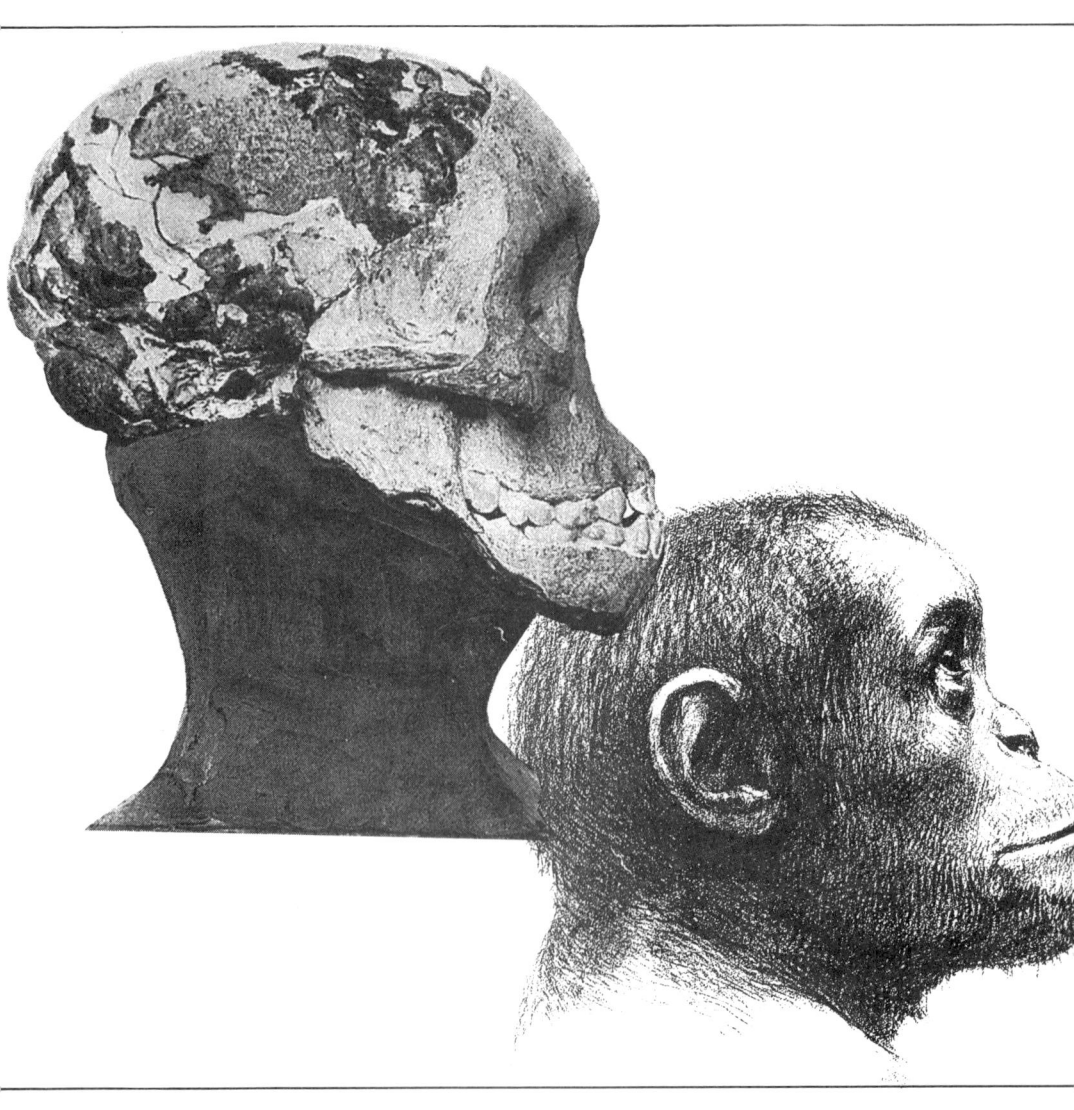

Raymond Dart's discovery of the Tuang baby skull (to the left, with a reconstruction drawing on the right) provided evidence of a previously unknown genus *of human ancestors.*

ape, of course, is not really stooping; it is assuming the natural posture of an ape. But if a human were to imitate the posture, he or she would have to stoop forward to do so.) In the tiny fossil skull, the *foramen magnum* was on the bottom, which seemed to indicate that the creature had stood upright. And no known species of ape stands upright, except for brief bursts of upright locomotion.

Dart decided that the skull represented a hitherto unknown genus of human ancestor that he named *Australopithecus,* or "southern ape," presumably because the fossil had been found in South Africa. This genus, Dart believed, was an earlier member of the family *Hominidae*. Specifically, the species to which the original owner of the skull had belonged was *Australopithecus africanus*. However, because this particular fossil was the skull of an infant, it inevitably became known to the general public as the *Taung baby.*

If Dart was correct, *Homo sapiens* and his close relatives *erectus* and *neanderthalensis* were no longer alone on their particular branch of the tree of life. There was a second genus in the family *Hominidae*. But Dart's discovery was largely rejected by other scientists of the period. Only in the 1940s did discoveries by the paleontologist Robert Broom convince others in the field that *Australopithecus* was a genuine member of the family *Hominidae* and not a prehistoric ape.

Among other things, Broom determined that there had been more than one species of *Australopithecus*. Some of the individuals that Broom discovered, which he called *Australopithecus robustus*, were massively built creatures compared to the relatively delicate *Australopithecus africanus* discovered by Dart.

Paleoanthropologists now refer to the delicately built members of the genus *Australopithecus* as the *gracile australopithecines*, and the more massively built members as the *robust australopithecines*. The robust australopithecines seem to have come along more recently in time than the gracile australopithecines, from which they were probably descended. Some paleoanthropologists have even suggested that the robust australopithecines should be put in a separate genus from the gracile ones. The name *Paranthropus* ("half-man") has been proposed for this genus, though it is not yet widely accepted.

Even as Dart was defending the validity of *Australopithecus*, a cave full of bones turned up in China that appeared to belong to a human ancestor. Popularly known as *Peking man* and given the scientific name of *Sinanthropus pekinensis* ("Chinese man from Peking"), these are now regarded as the remains of a late subspecies of *Homo erectus*.

In 1927, Canadian scientist Davidson Black announced important discoveries in this cave outside Peking, China. In the 80-foot (24-m) excavations, he uncovered fossils that came to be known as Peking man.

THE FOSSIL FINDS mentioned in this chapter were sufficient to prove Darwin's claim, at least within the scientific community, that humans had descended from a common ancestor with apes. But in a sense, these fossils are like the pieces of a jigsaw puzzle that have not yet been put together. Each new discovery only raised new questions: How ancient were these ancestors? How long ago did humans have a common ancestor with apes? Have the fossil remains of that ancestor been found? How intelligent were these ancient ancestors? Could they use tools? What came first, erect posture or large brains?

Perhaps most importantly, from the viewpoint of nonscientists confused by this seeming wealth of ancient species, are these questions: How are we related to these other species? Are they our direct ancestors on the tree of life? Or are they our cousins, our uncles and aunts?

In the next few chapters, we'll see how scientists have attempted to answer these questions, starting with their attempts to make sense of the human family tree and its position on the great tree of life.

PART TWO

THE COMMON ANCESTOR

FOUR

THE MISSING LINK

Human beings, as we have seen, belong to the family *Hominidae*. By convention, members of this family are known collectively as *hominids*. Although humans are the only hominids living on earth today, the discoveries that we reviewed in the last chapter told scientists that there have been other hominids in the past. Some of these, such as *Homo erectus*, are in the same genus, *Homo*, to which modern humans belong. One, *Homo sapiens neanderthalensis*, is merely a separate subspecies under our own species, *Homo sapiens*. The australopithecines, on the other hand, are a separate genus from the genus *Homo*. But they are considered hominids because they are in the family *Hominidae*.

The main characteristic that distinguishes hominids from other families is their erect posture.

Hominids, whether *Homo* or *Australopithecus*, are the only mammals that habitually walk on two legs, with their heads held directly above their hips.

The hominids, in turn, are members of a larger superfamily called the *Hominoidea*. This superfamily includes both hominids—human beings and our extinct ancestors—and the apes, such as the chimpanzees, the gorillas, and the orangutans. *Hominoidea* is the next node above *Hominidae* on our particular branch of the tree of life.

As we saw in the last chapter, nineteenth-century paleontologists had a deep interest in finding the "missing link," the common ancestor of humans and apes. The missing link would have been the progenitor of all the hominoids, the members of the superfamily *Hominoidea*.

The sheer wealth of ancient human ancestors discovered in the nineteenth and twentieth centuries took the spotlight off the search for the missing link. Who needs a missing link, after all, when so many fascinating links have already been found? But for a time in the mid-twentieth century, paleontologists genuinely believed that they had found something quite close to the common ancestor of humans and apes. And this belief led directly to one of the greatest controversies of modern paleoanthropology, one that was only resolved in the early 1980s and that also led scientists to a revolutionary new method of studying the evolutionary past. In this chapter, we'll look at the search for the common ancestor from the view-

The long-sought "missing link" would have been the common ancestor of human beings and the apes— the orangutan (above), chimpanzee, and gorilla.

point of scientists who study fossils. In the next chapter, we'll take a look at the search from the viewpoint of a new group of scientists called *molecular anthropologists.*

EVOLUTION HAS been in progress for at least four billion years. The ancestors of humans, apes, and everything else were self-replicating molecules in lakes and oceans. These self-replicating molecules eventually gave way to single-celled organisms, which were the ancestors of early multicelled organisms similar to modern sponges and jellyfish.

The next important node on the evolutionary branch leading to the hominoids were the fish. About 400 million years ago, a rather peculiar fish of a type we call lungfish moved onto dry land and "gave birth" evolutionarily to the amphibians, which in turn spawned the reptiles. And from the reptilian branch of the tree there shortly emerged a brand new branch that we call the mammals.

Paleontologists like to divide the history of the earth into periods of time called *eras.* The last 600 million years (the period from which the most fossils have been found) is divided into three great eras: the Paleozoic (old life), the Mesozoic (middle life), and the Cenozoic (recent life). The Cenozoic era is still in progress today.

Mammals evolved during the early part of the Mesozoic, about 240 million years ago. However, the entire Mesozoic era, which lasted until roughly 65 million years ago, was dominated by reptiles, including the reptiles that we now call dinosaurs.

The way in which an organism functions within its environment—the type of food it eats, the way it obtains that food, the way it reproduces, etc.—is called its *ecological niche*. It is a truism of evolutionary biology that no two species can occupy the same ecological niche in the same environment at the same time; the competition between them would be so fierce that one species would soon die off or be forced to move to another environment. Fortunately, there are many different ecological niches in any given environment, and thus many varieties of organism can live together within an environment.

During the Mesozoic, all of the ecological niches available to animals larger than, say, housecats were occupied by reptiles. All of them. These included the ecological niches that are today occupied by such animals as dogs, lions, elephants, eagles, porpoises, whales, and so forth.

Mammals, therefore, were forced to occupy only those ecological niches that, for one reason or other, the reptiles were not suited for. Thus, no Mesozoic mammal was larger than a housecat—few, actually, were larger than small rats—and most were nocturnal; that is, they slept during the day and searched for food at night.

In effect, the Mesozoic mammals lived under the feet of the dinosaurs (which at the time were the dominant land-based reptiles). This forced them to develop brains that were large and quick, at least by the standards of the time. Hunting at night also forced them to develop especially good eyesight.

The mammals of the Mesozoic included the distant ancestors of that group of mammals that today we call the *primates,* which includes both monkeys and apes. These early primates were insectivores (they lived by eating insects) and resembled the modern shrew.

At the end of the Mesozoic, most of the reptiles became extinct. No one is quite sure why, though a current favorite theory suggests that it was the result of a collision between the earth and an asteroid or a comet, which raised great clouds of dust and thus temporarily but violently altered the world climate. The disappearance of the reptiles opened a large number of ecological niches, and the mammals rapidly evolved to fill them. Thus, if the Mesozoic can be considered (as it usually is) the Age of Reptiles, the following era—the Cenozoic—can fairly be regarded as the Age of Mammals. Not only do mammals fill most of the Cenozoic niches for large land animals, but there are also mammals that live in the sea (i.e., whales and dolphins) and mammals that fly in the air (i.e., bats).

In the early Cenozoic, many of the early primates evolved into an arboreal, or tree-dwelling, niche, probably because insects could be found among the fruits and flowers on the tree branches. Once in the trees, however, the primates evolved to

The ancestral primate was probably a small, nocturnal animal that lived in the trees and fed on insects, much like this tree shrew of today.

eat the fruits and nuts that they found there. These tree-dwelling primates became primitive forms of monkeys known as *prosimians*.

Life in the trees encouraged these primates to evolve quick reflexes and even better vision; a primate that could see and grab a branch in a hurry was less likely to fall to a quick death on the ground below.

Specifically, the early primates developed color vision and stereoscopic, or 3-D, vision. Some experts have suggested that these developments, in turn, encouraged the primates to develop more advanced brains, to sort out all this sensory data. Even today, almost half of the human brain is devoted to processing the information that we receive through our eyes, reflecting the importance of vision to the evolution of brainy primates.

Tree dwelling also encouraged the evolution of hands capable of great dexterity. Originally, these more advanced hands were intended for such tasks as grabbing branches and picking fruit, but millions of years later they would prove useful for the manipulation of tools.

Even as the primates were evolving, so was the earth itself evolving. The continents of the earth were drifting apart, slowly forming the familiar arrangement that we see on modern maps. At the time the prosimians were evolving, the continents that we today call South America and Africa were joined together into one large continent. This continent subsequently broke apart, carrying the ancestors of early monkeys toward opposite sides of the globe. The monkeys on the African continent evolved into what we now call the *Old World monkeys*. The monkeys on the South American continent evolved into what we call the *New World monkeys*.

The ancestral primates of the Old World eventually gave evolutionary birth to the animals that we today refer to as the apes. There are two main branches of the ape family, one of which lives in Africa and the other of which lives in Asia. The African apes are the gorilla and the chimpanzee. The Asian apes include the orangutan and the gibbon.

It has been known since at least Linnaeus's day that the African apes are closer to human beings on the tree of life than are the Asian apes. In fact, they are closer to human beings on the tree of life than any other organism alive today.

In evolutionary terms, this means that we shared a common ancestor more recently with the African apes than with the Asian apes or with any other species. Thus, the African apes are the descendants of those apes that evolved before the hominids.

And that brings us to the question with which we began this chapter: What species of creature represented the common ancestor of humans and the African apes? What is the missing link?

IN 1934, paleontologist G. Edward Lewis discovered the fossil remains of a prehistoric apelike creature in the Siwalik Hills of India. He called it

*African apes include the gorilla (above);
Asian apes include the orangutan (opposite).*

Ramapithecus, or "Rama's ape," after a mythical Indian prince. Despite the name, Lewis believed that this fossil was in fact not that of an ape at all but of a very early hominid, perhaps the earliest hominid of all. *Ramapithecus,* according to Lewis, was not quite the missing link, but it was as close to it as any fossil ever found and perhaps as close as any that was ever likely to be found. He believed, in fact, that *Ramapithecus* was very nearly

the direct descendant of that common ancestor, the first node on the branch of the tree of life that led to *Homo sapiens.*

Lewis, however, was pretty much alone in that idea. His theory was soon shot down by one of the leading paleontologists of the period, Alex Hrdlicka. And though many scientists now agree that Hrdlicka's rejection of the theory was not particularly well reasoned and was probably based on private motives of Hrdlicka's own, the theory nonetheless failed to find acceptance in the scientific community.

But nearly thirty years later, the theory was revived by Elwyn Simons of Yale University. In 1961, Simons published a paper entitled "The Phyletic Position of *Ramapithecus,*" in which he argued that *Ramapithecus* was the earliest known (and perhaps the earliest) hominid.

The fossil remains of *Ramapithecus* are scanty, as are the remains of most fossil organisms. In fact, the remains on which both Lewis and Simons based their conclusions were primarily the two broken halves of a lower jaw.

It may seem remarkable that scientists can base important theories about extinct organisms on such seemingly small amounts of evidence. However, small amounts of evidence are frequently all that is available, and thus these scientists have had to learn to perform amazing feats of deduction to come up with their theories. Surprisingly, these theories often hold up quite well in the face of additional evidence.

In the case of *Ramapithecus,* the argument that this fossil represented an early hominid was based almost entirely on the nature of the jaw. For one thing, *Ramapithecus* had small canine teeth. It had been suggested by Darwin himself that as early human ancestors developed increasingly dexterous use of their hands, their teeth would have grown smaller, since there was less need to manipulate objects by grasping them in their jaws.

More importantly, the jaw of *Ramapithecus* was thought by Simons to have a shape that more closely resembled that of a human than an ape. The lower jaw of an ape is nearly rectangular, with the molars arranged in parallel lines on the sides and the front teeth at right angles to the molars. In humans, on the other hand, the jaw is rounded almost into a semicircle. Simons carefully reconstructed the broken jaw of *Ramapithecus* and decided that it was too rounded to be the jaw of an ape. Hence, *Ramapithecus* was deemed a hominid.

The idea caught on. Soon, Simons was joined by a colleague named David Pilbeam, and the two became the greatest champions of the theory that *Ramapithecus* was a very early hominid. This theory was soon accepted by the rest of the anthropology community.

Perhaps the most important fact about this theory is that it told scientists that hominids had been around for a long time. *Ramapithecus* was believed to have lived between about 15 and 20 million years ago. Compared to the 4.5-billion-year age of the earth, this may not seem like a very long time, but remember that it does represent a substantial fraction of the time that primates have been evolving. It also meant that, although apes may be the closest relatives of humans on the tree of life, they really aren't all that close. Chimps and gorillas, for instance, would be much more closely related to each other than either is to humans. The common ancestor of chimps and gorillas would have arrived on the scene much more recently than the common ancestor of humans and the African apes.

This seemed to fit well with common sense. To the human eye, chimps and gorillas resemble each other far more than either does a human being. They are both covered with fur, for one thing, and humans are relatively naked (which makes us unique among primates). Chimps and gorillas both propel themselves along the ground through a form of locomotion known as knuckle walking, which involves leaning forward in a typical primate posture and propelling oneself along with knuckles scraping the ground. Humans, on the other hand, walk entirely erect, using the hands primarily for manipulating and carrying objects. And neither chimps nor gorillas have developed either language or advanced technology, as have humans. Clearly, there was considerable evolutionary distance between humans and apes.

Or was there? Could this merely be a kind of human chauvinism, a species snobbishness? Are we really as far advanced beyond the apes as we would like to believe we are? Does *Ramapithecus* really prove that it has been at least 15 million years (or, as some scientists began to suggest, as much as 30 million years) since humans branched off from the hominoid limb of the tree?

In the late 1960s, new evidence appeared that suggested *Ramapithecus* was not a hominid at all. In fact, *Ramapithecus* was almost certainly not an ancestor of either humans *or* African apes, but an ancestor of the less sophisticated Asian apes. Yet this evidence came entirely from outside the field of paleoanthropology, and it was immediately and soundly rejected by the paleoanthropologists.

FIVE

THE MOLECULAR LINK

In the years before the printing press was invented, books were copied laboriously by hand, usually by dedicated monks. Because of the time and expense involved in producing them, only a very few books, such as the Bible, were distributed to large numbers of people.

Imagine that you are a scientist who studies such books. You amass as many copies of one of these books as you can afford. (Such books are now extremely rare collectors' items, but we'll assume you have the budget of a major university behind you.)

Suppose further that the books have no dates on them, so you don't know when each one was transcribed. Is it possible to arrange the books in chronological order, according to the date each copy was made?

In a sense it is. The reason is that the copying monks sometimes made mistakes. And when another monk made a copy of the copy that contained the mistake, he would probably copy the mistake made by the earlier monk and quite possibly add a mistake or two of his own. Thus, we can put the books in chronological order by putting them in the order of increasing errors.

Let's say that you have three copies of the book. We'll call them Copy A, Copy B, and Copy C. Copy A is flawless, with no mistakes at all. Copy B contains a single mistake. And Copy C contains that same mistake plus a brand-new mistake. You could guess, with a fair chance of being correct, that Copy A was the oldest of the copies, Copy B was the next oldest, and Copy C was the newest. The reason, of course, is that errors would tend to accumulate as time went by. The older the book, the fewer errors it would have.

Now suppose that you find a fourth copy of the book—call it Copy D. Copy D contains the same error as Copy B, plus a second error. But this second error is not the same as the second error in Copy C. Where does this new copy fit into the chronology?

Probably, Copy D was copied from Copy B (or from a copy of Copy B), just as Copy C was. It is, in effect, a cousin of Copy D. The copying process must have "branched" at this point, and Copy B represents the node where the branching took place. We can even draw a "family tree" for the three copies of the book, as illustrated in Figure 3.

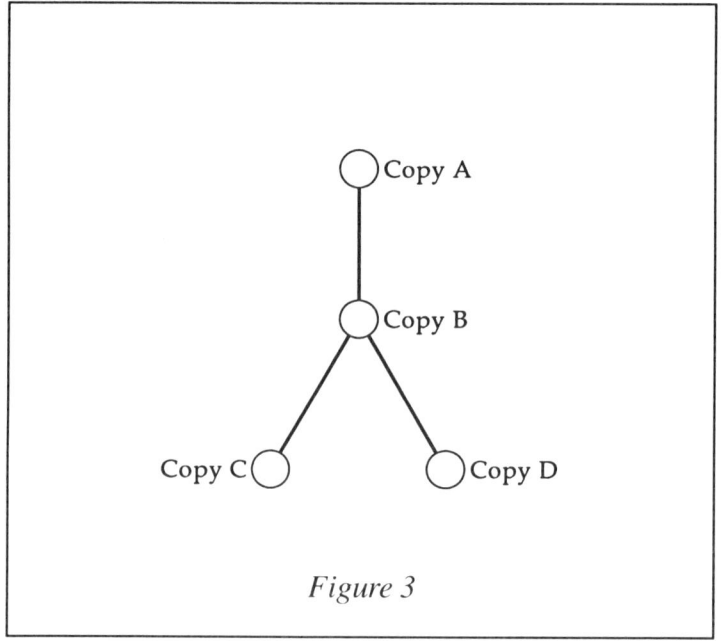

Figure 3

It may seem strange to think of books as having a family, of one book descending from another book as a child descends from a parent or one species descends from another species. And yet the processes are very similar, if we assume that the books are copied from one another in the fashion described above.

In the same way, when one organism produces another organism, its chromosomes are carefully copied and passed on to its offspring. As we saw in Chapter Two, the DNA molecules in chromosomes can be looked on as a kind of book, written in an alphabet made up of special molecules.

Just as the monks who transcribed books occasionally made errors in copying, so living organisms can make mistakes in copying their genes. These mistakes are the mutations that we spoke of earlier.

So, in theory, it should be possible to compare the DNA molecules of living organisms and deduce which organisms are descended from common ancestors (by the similarities in their "errors"). If we are very, very clever, we might just be able to deduce how recently that common ancestor lived on earth.

This idea is not new. One of the first scientists to apply it to the study of evolutionary relationships was the British biologist Henry Falkiner Nuttall, near the end of the nineteenth century. This was before the genetic code had been deciphered, even before the chromosomes had been identified as the location of the genetic information. Nuttall studied substances called antibodies found in the blood of most animals and found that the antibodies in humans more closely resembled the antibodies in the Old World primates than those in the blood of New World primates or the blood of any other animal. Thus, the Old World primates were our closest relatives. This was not surprising news to paleoanthropologists, but it was an encouraging corroboration of what they had already deduced from other evidence.

The idea of comparing species chemically languished for more than half a century after Nuttall, until the early 1960s. By this time, the genetic code was well understood by molecular biologists and biochemists—scientists who study the workings of living organisms on the levels of atoms and molecules—and the chromosomes were known to be the site of genetic information.

In 1967, two biochemists, Vincent Sarich and Allan Wilson of the University of California at Berkeley, used sophisticated techniques to compare certain molecules from the bodies of human beings with similar molecules from the bodies of apes. The way in which these molecules are constructed, from substances known as proteins, is determined by the information in the genes; in fact, the genes can be looked on as a set of recipes for making protein molecules. Thus, comparing protein molecules between humans and apes is a good way of determining how many differences there are between the genes of the two species.

What Sarich and Wilson learned from this was surprising. Not only are the African apes our closest relatives (this had been known, or at least guessed at, for more than a century), but our genes are 99 percent identical! That is, the differences between humans and certain apes are contained in only 1 percent of our genes. These apes are even closer cousins than anyone had previously realized.

But Sarich and Wilson went even further than this. Not only did they determine how closely re-

By comparing protein molecules between humans and apes, researchers learned that the African apes, including gorillas, are our closest relatives.

lated humans and apes are by comparing their protein molecules (and thus their genes), they also determined how long it has been since the common ancestor of the two species existed.

How could they do this? According to Sarich and Wilson, the molecules of the body can be looked on as a kind of clock, a molecular clock. Remember that the genes are composed of nothing more than the information in the chromosomes, or DNA molecules, and this information is written in an alphabet made up of four smaller molecules. Every now and then, when a chromosome is being copied, an error is made, and the information is incorrectly transcribed. An error occurs.

Usually, these errors take the form of *point mutations;* that is, a single "letter" of the molecular alphabet is incorrectly transcribed, so that an A becomes a C, a T becomes a G, or some similar error. Surprisingly, such errors usually have no effect on the creature that carries them in its genes. Thus, we call these errors *neutral mutations,* mutations that make no particular difference to the species.

Nonetheless, these mutations are real and can be spotted or at least counted by using sophisticated comparison techniques on a pair of chromosome molecules. (It's easier to tell how *many* differences there are between two molecules than to determine exactly what the differences *are.*)

Point mutations occur at random. There is no way to predict that one is going to happen. But the average number of point mutations that will occur in the chromosomes of a species over a long period of time is very predictable. And thus, the accumulation of mutations is rather like a clock. By counting the number of differences between two protein molecules that were once identical, we can measure how long it has been since they belonged to the same species.

Since apes and human beings descended from a common ancestor, there was presumably a time when their chromosomes were identical. (Actually, no two organisms on earth have identical chromosomes, except for identical twins, triplets, etc. However, there are ways in which biochemists such as Sarich and Wilson can compensate for this fact.) By counting the differences that exist in the chromosomes of the species today, we can tell how long it has been since the species became separate, given that mutations occur at a steady rate.

This is exactly what Sarich and Wilson did. Their conclusion was that humans and apes belonged to a single species approximately five million years ago.

When this result was announced, Sarich and Wilson expected that the information would revolutionize the study of prehistoric hominids. But what actually happened was that paleoanthropologists simply refused to believe it at all.

In the late 1960s, paleoanthropologists generally agreed that apes and humans had not been a single species—that is, shared a common ancestor—for at least 20 million and more likely 30 million years. The proof was that *Ramapithecus*, which was widely regarded as an early hominid, had lived at least 15 million years ago, and so the common ancestor must have lived even earlier.

And yet, here were these two upstart biochemists claiming that this figure was several times too big. Paleoanthropologists were appalled!

Elwyn Simons, the main proponent of the theory that *Ramapithecus* was an early hominid, wrote in 1968 that "[i]f the . . . dates of divergence devised by Sarich are correct, the paleontologists have not yet found a single fossil related to the ancestry of any living primate . . . I find this impossible to believe. It is not presently acceptable that *Australopithecus* sprang half-blown five million years ago, as Minerva did from Jupiter, from the head of a chimpanzee or gorilla."[1]

Sarich did not help the situation any when he made the widely publicized remark that "[o]ne no longer has the option of considering a fossil specimen older than about eight million years a hominid no matter what it looks like."[2] It was bad enough that the new theory contradicted the conventional beliefs of paleoanthropologists; now Sarich was telling them how to interpret fossils!

The main argument that the paleoanthropology community mounted against Sarich and Wilson was that the molecular clock did not keep good time. There was no way that random events such as point mutations on a chromosome could be directly correlated with the passage of time.

And yet, Sarich and Wilson had performed

meticulous studies to assure that the molecular clock was correct. For instance, they had compared the genes of other species for which the evolutionary tree was well established (by the fossil record), and the clock worked every time. It was amazingly accurate.

Unfortunately, when Sarich and Wilson originally obtained their results, they wrote them up in three scientific papers, two of which detailed their findings and one of which explained how they calibrated the molecular clock. *Science* magazine, which is widely read throughout the scientific community, published the two articles about the results but refused to publish the article about the clock, claiming that it didn't really contain any new information. The clock article was eventually published in another magazine but was not widely read. Thus, many paleoanthropologists may not have been aware of the amount of work that had gone into calibrating the molecular clock.

It took fifteen years—from 1967 to 1982—before Sarich and Wilson's findings were accepted by paleoanthropologists. And the reason that the findings eventually were accepted had nothing to do with Sarich and Wilson's work. It had to do with the discovery of new fossils.

In 1980 and again in 1982, fossil remains of an apelike creature known as *Sivapithecus* were discovered. *Sivapithecus* was long known to be a close relative of *Ramapithecus*. But these new *Sivapithecus* fossils were of better quality than any previously discovered, and it was clear from the structure of the bones that *Sivapithecus* was also a close relative of the modern orangutan.

How could this be? *Ramapithecus* was regarded as a hominid and therefore a closer relative of the African apes, such as the chimp and the gorilla, than of the Asian apes, such as the orangutan. Yet, if *Ramapithecus* were a close relative of *Sivapithecus*, then it must be more closely related to the Asian apes and could not possibly be a hominid.

The only solution was to accept that *Ramapithecus* was not an early hominid at all but an early relative of the Asian apes. And this removed the barrier to accepting Sarich and Wilson's figure of five million years (give or take a million) for the common ancestor of apes and humans.

Of course, that common ancestor has still not been found. But, as we shall see in the next two chapters, it is nonetheless possible that we have discovered a fossil representing just about the earliest hominid that lived on earth, dating back the better part of four million years.

THE ACCEPTANCE of Sarich and Wilson's date for the common ancestor means, of course, that paleoanthropologists are now willing to regard molecular dating as an important tool in the study of evolution in general, not just the evolution of humans in particular. When fed into a computer, molecular information can be used for studying the branchings of the fossil tree of life, even in the absence of clear fossil evidence.

Computers, in fact, are well adapted for working with data structures shaped like trees. In fact, it is rather amazing how many computer applications such tree structures can have. For instance, computer programs that play chess often do so by regarding the game as a large tree graph, where each node represents a possible position of the board and each branch represents a move that gets the board from that position to another position. The computer chooses which move to make by searching the branches of the tree to see which branches of which nodes lead to a position of maximum advantage over the computer's opponent.

A computer given information about the chromosomes of several organisms can use that information to draw a family tree that shows the evolutionary relationship between those organisms. Linnaeus would have been astounded by such a computer, but he probably would have understood what it was doing, at least in principle.

Figure 4 shows such a graph for several species. The nodes of the graph represent the points at which common ancestors of these species existed. Thus, we can see, for instance, that humans are more closely related to chimpanzees than to gibbons, and that dogs are more closely related to foxes than to tigers. And who would have thought that humans are more closely related to herring than to squid!

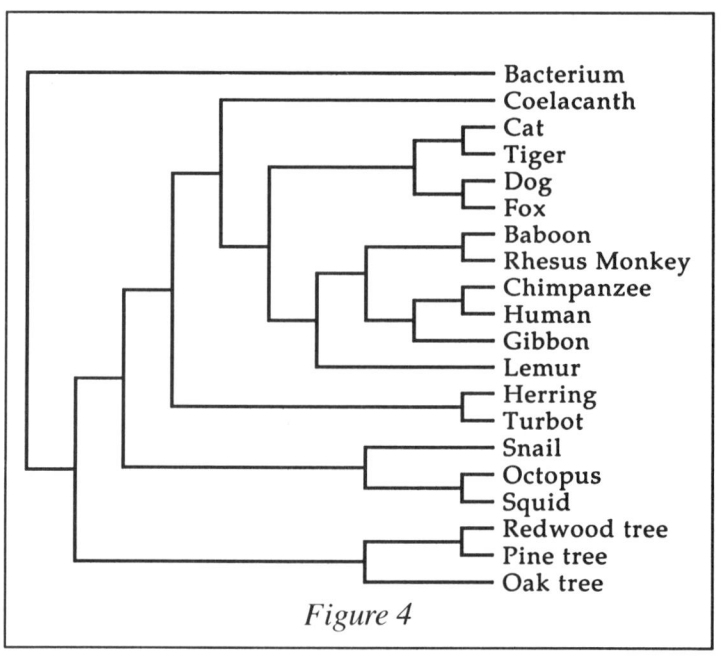

Figure 4

Unfortunately, the more species that we put into such a graph, the longer it takes the computer to produce it. Zoologist Richard Dawkins estimates that there are 8,200,794,637,891,559,375 possible tree graphs representing the relationships between twenty different animals.[3] A computer generating such a graph would have to try each possible tree in order to find the right one, which might take more time than any existing computer has to devote to the problem, something on the order of, say, ten billion years or thereabout. Fortunately, there are shortcut methods that the computer can use to find an *approximately* correct tree graph in a much shorter period of time.

There are things that molecular clocks cannot tell us, of course. For instance, they can tell us nothing about the relationships between species that are now extinct, simply because the chromosomes of these species are no longer available for study. Nonetheless, as we shall see at the end of Chapter Seven, the molecular anthropologists can tell us some fascinating things about the origin not of just the hominids but also of our very own subspecies, *Homo sapiens sapiens*.

PART THREE

THE FAMILY TREE

SIX

THE LEAKEY BUSH

July 17, 1959. Paleoanthropologist Louis Leakey lay feverish with malaria in his tent at Olduvai Gorge, in the African nation now known as Tanzania. Suddenly, he heard his wife, Mary, outside the tent, excitedly declaring that she had just discovered a hominid fossil. Though ill, Leakey jumped out of bed and raced to the site, where his wife showed him the skull that she had found.

Leakey's smile vanished. "Why it's nothing but a [expletive] robust australopithecine!" he growled.[4]

This, at least, is one version of the story. His wife remembers a somewhat milder version of Leakey's exclamation: "Oh, dear," she recalls him saying. "I think it's an australopithecine."[5]

Whatever Leakey said, it's obvious that he was less than thrilled that his wife's momentous discovery had turned out to belong to the *Australopithecus* genus. And to understand Leakey's disappointment, it's necessary to understand a little about Louis Leakey.

Leakey was almost certainly the best-known paleoanthropologist of the mid-twentieth century. Without a doubt, he was the most flamboyant. He was a larger-than-life figure with firm convictions and no hesitation about voicing them.

One of these convictions was that the genus *Homo* was much older than most other paleoanthropologists believed, that the big brains and erect posture that mark our species have been around for quite a few million years. And Leakey believed that eventually he would find the fossils that would prove the antiquity of the genus *Homo*.

But he also believed that *Australopithecus* was irrelevant to the history of humanity, or at best it was a minor sideline to the main story. According to Leakey, *Australopithecus* was a distant cousin of early humans, not a direct relative. For one thing, *Australopithecus* was too small-brained to be a relative of the brainy prehistoric humans. And although it was inevitable that the genus *Homo* had descended from smaller-brained ancestors, according to Leakey those ancestors could not possibly be the australopithecines, because *Australopithecus* had been around much too recently. Since some *Australopithecus* fossils had been found to date back barely more than two million years, they could not possibly be ancestors of the

Louis Leakey devoted his life to finding the proof that Africa was the cradle of humankind.

genus *Homo*—or else Leakey would be wrong about the age of that genus.

Thus, Leakey was less than enthusiastic about his wife's having discovered a fossil of *Australopithecus*, which he regarded as relatively unimportant to the development of the human race. But what really bothered Leakey about this particular *Australopithecus* fossil was that it had been found in direct association with a large number of ancient stone tools.

Tools are often more common in fossil beds than are hominid skeletons. They are more resistant to the ravages of time and less likely to be destroyed by hungry predators. Leakey had been finding ancient tools since he was a boy growing up in Kenya.

The ability to manufacture a variety of tools has traditionally been regarded as one of the abilities that sets human beings apart from other animals, including our nearest ape relatives. (In recent years, it has been discovered that chimpanzees also make limited use of tools.) And it is our distinctively large brains that allow us to make tools. Tools are a sign of intelligence.

Yet the *Australopithecus* fossil discovered by Mary Leakey was found in the vicinity of a large cache of tools. Was it possible that the small-brained *Australopithecus*, generally regarded as no more intelligent than a modern ape, could have manufactured tools after all?

Louis Leakey could not accept this conclusion. *Australopithecus* was not a toolmaker. Period.

Olduvai Gorge, on the southern edge of Tanzania's Serengeti Plain, is where Louis and Mary Leakey spent over thirty years in search of human ancestors.

Yet in the absence of further evidence, it seemed clear that Mary Leakey's fossil hominid was the maker of the tools found at Olduvai Gorge.

Leakey was thus on the horns of a dilemma, and he did the only thing that he felt he could do under the circumstances. He declared that the fossil belonged to a new genus of hominid, one that had never been discovered previously. It was neither *Australopithecus* nor *Homo;* it was (said Leakey) *Zinjanthropus boisei.*

Roughly translated, this name meant Boise's East African man. (Charles Boise was the major financial backer of Leakey's expedition to Olduvai Gorge.) The fossil soon became known around the world by the affectionate nickname "Zinj." (It was also known as Nutcracker man, because of its unusually broad teeth.)

According to Leakey, Zinj was a hitherto unknown toolmaker that was a direct ancestor of neither *Homo* nor *Australopithecus* but was a cousin to both. Obviously, *Zinjanthropus* had subsequently become extinct and thus had left no living descendants.

Then, less than a year and a half later, a second fossil find occurred at Olduvai Gorge that eclipsed the discovery of Zinj. Leakey's son Jonathan found several portions of a hominid skull not far from where Zinj had turned up. And this skull clearly belonged not to *Australopithecus* or *Zinjanthropus* but to the genus *Homo.*

Leakey was ecstatic. Not only did this skull belong to the genus *Homo,* but it appeared to be the oldest representative of the genus yet discovered, even older than *Homo erectus*, thus helping to prove Leakey's contention that the genus *Homo* was older than anyone thought. In fact, the skull was eventually shown to be about two million years old. After several years of study and deliberation, Leakey announced on April 4, 1964, that this skull belonged to a hitherto unknown species that he called *Homo habilis*, which roughly translates from Latin as "handy man."

Ironically, the discovery of *Homo habilis* freed Leakey from the dilemma presented earlier by the discovery of Zinj. Now it was clear that it was *Homo habilis* and not *Zinjanthropus* that had manufactured the tools found in the vicinity. As there was no longer any reason not to regard Zinj as a robust *Australopithecus*—in fact, far from being the maker of the tools at the Gorge, Zinj had probably been killed and eaten by *Homo habilis* with those same tools—the species was quietly renamed *Australopithecus boisei*. Zinj was "a [expletive] robust australopithecine" after all.

THIS ILLUSTRATES one of the major problems of modern paleoanthropology—finding a sensible family tree that connects all the known species of early human beings and other hominids. Which fossil species are direct ancestors of one another? Which are collateral relatives (i.e., cousins)? Which are direct ancestors of *Homo sapiens sapiens*?

In the early days of paleoanthropology, as sketched in Chapter Three, it was assumed (for lack of better evidence) that all prehistoric hominids were direct ancestors of modern human beings. This idea is known formally as the *single-species hypothesis,* and it says that only one species of human ancestor lived on earth at any one time.

This makes a certain degree of sense, when we consider that only one species can occupy any given ecological niche within an environment at one time. If we assume that all hominids occupied the same ecological niche and lived in more or less the same environment, then only one hominid could have lived on earth during any historical period.

But as our knowledge of prehistoric hominids grew, it became increasingly evident that the single-species hypothesis was an oversimplification, to say the least. Certain species of *Australopithecus,* for instance, clearly lived at the same time as certain species of *Homo.* The coexistence of *Australopithecus boisei* and *Homo habilis* at Olduvai Gorge is but one example.

Clearly, then, at least one species of *Australopithecus* was a cousin of at least one species of the genus *Homo.* How is this possible if they lived in the same environment? Apparently, they occupied different ecological niches. *Australopithecus boisei* was, it seems, a vegetarian, using its large teeth to crunch leaves. *Homo habilis,* on the other hand, must have had more carnivorous habits, eating, among other things, *Australopithecus boisei.* (This seems rather cannibalistic to modern sensibilities, but they were, after all, a different species.)

The single-species hypothesis has survived into modern times but in modified form. One recent version of the hypothesis is illustrated in the hominid family tree in Figure 5.

In this version, *Australopithecus africanus* (the so-called "gracile australopithecine") is the direct ancestor of all later hominids, including human beings. *Australopithecus robustus*, on the other hand, is only a cousin of the genus *Homo*, which is represented by *Homo habilis* (the "grandparent" of modern humans), *Homo erectus* (the "parent" of modern humans), and *Homo sapiens* (modern humans).

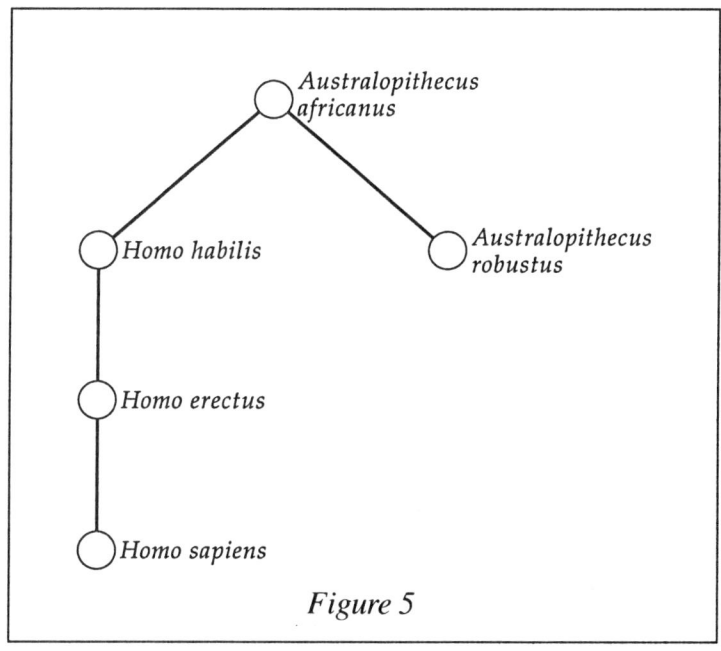

Figure 5

Louis Leakey was adamantly opposed to this version of the hominid family tree. Leakey liked to say that the tree was not a tree at all but a bush, which is to say that instead of one long branch, it was made up of a lot of relatively short branches running side by side. To Leakey, most species were cousins.

In fact, Leakey did not even believe that *Homo erectus* was an ancestor of human beings, though most paleoanthropologists disagree with him on this point. Instead, he believed that modern *Homo sapiens* had descended directly from *Homo habilis.* And who had *Homo habilis* descended from? Leakey did not know but assumed that an appropriate fossil would eventually be found. Possibly, *habilis* itself was a very long-lived species, one that had descended directly from the common ancestor of humans and apes.

Leakey's family "bush" for the hominids might have looked rather like the one in Figure 6, with almost every hominid on its own branch, except for *Homo habilis* as the direct ancestor of *Homo sapiens* and *Australopithecus africanus* as the direct ancestor of both *robustus* and *boisei*. Other bushes could also be constructed around Leakey's view of the hominid family.

Which is the correct family tree? At this date, it's still impossible to say. As new hominid fossils are discovered, more information becomes available for tree making, but it's possible that a definitive family tree may never be constructed. Unlike other sciences—physics, say, or astronomy—the ultimate questions of paleoanthropology may not

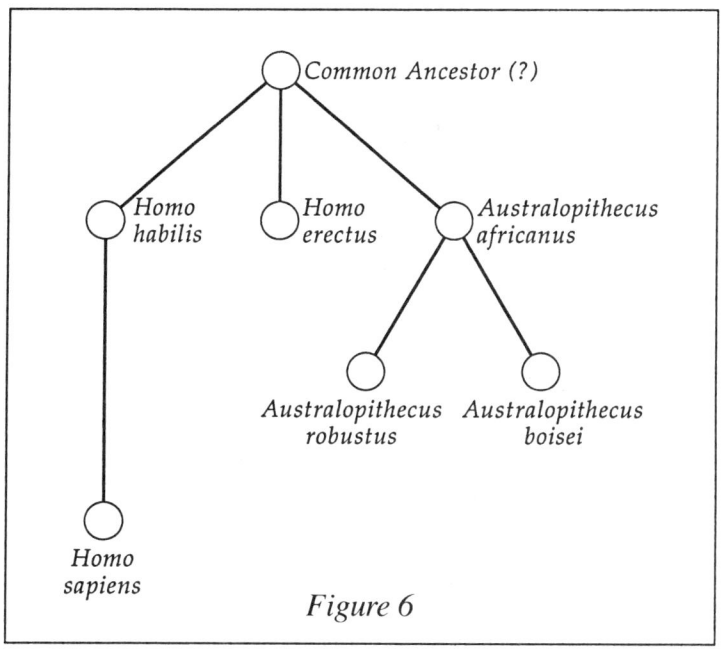

Figure 6

be answerable. That is to say, the "correct" answers lie deep in the past. The fossils that might answer these questions might not exist, or if they do exist, they might never be found. Barring the invention of a time machine, the proper construction of the hominid family tree may never be known.

Still, much of the controversy in recent anthropology has stemmed from attempts to construct such a tree. And, as time goes by, more and more information is available for us to put into that construction.

THE SKULL had been shattered into more than three hundred pieces by millions of years of pressure. It lay inside the earth like the fragments of Humpty-Dumpty, waiting for all the king's horses and all the king's men to try and put it back together again.

Instead, it was found by a team of paleoanthropologists under the general direction of Richard Leakey, son of Louis. All considered, that was probably for the best.

The skull was discovered in August 1972 in the general vicinity of Lake Turkana in northern Kenya, where Richard has maintained a permanent base for fossil hunters since the late 1960s. Even by 1972, the site had become famous for the sheer number of hominid fossils found there, including an excellent specimen of *Australopithecus boisei* and a mysterious fragment of skull that might have been *Homo habilis*.

The new skull was the most exciting find yet. Even though shattered into a large number of pieces, it looked as though it might be unusually complete and quite possibly it belonged to the genus *Homo*. If so, it may have been the oldest skull of that genus yet discovered.

Richard's wife, Meave Leakey, began the tedious task of putting the skull back together, as though it were an extremely complicated three-dimensional jigsaw puzzle. The task took her roughly a year to complete, but the results were well worth it.

The skull, known then and now as *Skull 1470*, was a nearly complete skull of *Homo habilis*. (Oddly, the skull has never been given a popular nickname, as major fossils of human ancestors

Meave and Richard Leakey display Skull 1470 and a hominid thighbone found near Lake Turkana in 1972.

usually are.) And there was evidence that the skull might be very old, perhaps 2.6 million years old, making it the oldest skull of the genus *Homo* ever discovered.

This tended to confirm the hominid family tree preferred by Richard's father, Louis. If the genus *Homo* were this old, then surely it had not descended from *Australopithecus*, as other scientists had suggested. And surely large-brained human ancestors had been around longer than anyone had suggested.

But the estimates of the age of Skull 1470 were almost immediately controversial. And the controversy stemmed from a geological feature known as the KBS tuff.

THE DATING of fossils is a tricky business. The earliest paleontologists, at the beginning of the nineteenth century, made no attempt to date fossils at all; rather, they identified fossils by the strata (layers of earth) in which they were found, on the assumption that the lower strata represented older fossils than the higher strata. In other words, the deeper in the earth a fossil is buried, the older it must be.

Thus, the early paleontologists assigned only relative dates to fossils rather than absolute ones. They could say that fossil A was older than fossil B or that fossil C was the same age as fossil D, based on the strata in which these fossils were found, but they couldn't date any of the fossils in years.

In recent decades, this situation has changed, with the introduction of *radioactive dating*.

Physicists learned in the early part of this century that certain substances are inherently unstable. The atoms of which these substances are made tend to break apart into other types of atoms over time. Because they release tiny particles and waves called radiation as they break apart, we say that these substances are *radioactive*.

The time that it takes a radioactive substance to decay—that is, change—into another substance can be measured very precisely, rather the way that Sarich and Wilson timed the mutations that take place in genes. By examining a fossil or rock that was "born" with radioactive substances inside of it, we can tell how old that fossil or rock is by determining how much of the radioactive substance has decayed over the years. In this way, we can date the age of the fossil or rock much as Sarich and Wilson dated the age of their molecules.

For recent fossils—those younger than about 50,000 years—a method of radioactive dating called *carbon-14 dating* can be used. This measures the amount of a radioactive substance called carbon-14 in the fossil, to see how much of it has decayed. Recent fossils can be dated quite accurately with this method.

Unfortunately, fossils as old as the *Homo habilis* skull found at Lake Turkana are too old to be dated with this method; the carbon-14 in these fossils has almost completely decayed. Thus, a second dating method, known as *potassium/argon dating*, must be used.

This method of dating measures the amount of the radioactive substance potassium-40 that has decayed into the gas argon-40. Unfortunately, it cannot be used to date fossils directly. In general, it is used only to date deposits of volcanic rock.

How, then, does this help us to measure the age of fossils? We said a moment ago that we can measure the relative age of fossils by noting which strata they are buried in. If we are lucky enough to find a fossil that is buried in a layer of volcanic rock, then we can also assign an absolute date to it. This is because we can be reasonably sure that the fossil is the same age as the volcanic rock in which it is buried, and we can measure the age of the volcanic rock using potassium/argon dating.

And it's not really even necessary that the fossil be in a layer of volcanic rock. If the fossil is directly below a layer of volcanic rock, for example, we can assume that it is slightly older than the rock, and if it is above volcanic rock, we can assume that it is slightly younger. If it is directly between two layers of volcanic rock, then it can be assumed to have an age between the two. And so on.

Layers of volcanic rock laid down in past ages are known to geologists as *tuffs*. Skull 1470 was found just below a layer of volcanic rock known as the KBS tuff. (KBS stands for "Kay Behrensmeyer Site," after the first geologist to study it.) Potassium/argon dating indicated that the age of the KBS tuff was approximately 2.6 million years. Skull 1470 was therefore given a date of approximately 2.6 million years by Richard Leakey.

Other paleoanthropologists didn't agree. In part, this stemmed from a rejection of the Leakey family's belief that the genus *Homo* was several million years old; many scientists believed that the genus was more recent than this, that *Homo* had descended from *Australopithecus africanus*. But there was also a more serious side to the disagreement. There were genuine reasons for disputing the date that had been assigned to the KBS tuff.

Specifically, the strata beneath the KBS tuff contained fossil animals—in particular, fossil pigs—that were nearly identical to fossil animals that had been discovered at other locations. But when found at other locations, these animals had been substantially younger than 2.6 million years, which would seem to indicate that the date assigned to the tuff was incorrect.

This controversy raged for several years, into the late 1970s, with Leakey maintaining the accuracy of the age assigned to the KBS tuff and the 2.6-million-year age that he had ascribed to Skull 1470.

In the end, Leakey was proved wrong. The KBS tuff was dated again, and the original date turned out to be incorrect. In fact, the tuff was slightly less than two million years old, making Skull 1470 roughly two million years old.

Although this is still an ancient specimen of the genus *Homo*, it is no more ancient than the *Homo habilis* specimen discovered at Olduvai Gorge in 1960 by Jonathan Leakey. It did nothing to increase estimates of the age of the genus. In addition, it still left open the possibility that the genus *Homo* had descended from *Australopithecus*, despite the impassioned belief of the Leakeys to the contrary.

Even as the controversy about Skull 1470 and the KBS tuff raged on, another series of fossil finds were taking place in nearby Ethiopia that were destined to cause yet another drastic alteration in the way in which paleoanthropologists viewed the hominid family tree. We'll look at those finds next.

SEVEN

THE OLDEST HOMINID?

Young paleoanthropologist Donald Johanson was up and out early on the morning of November 30, 1974, at his fossil dig in the Hadar region of Ethiopia. Johanson was showing a colleague, Tom Gray, the location of a section of the dig called Locality 162. He was in a good mood and later claimed that he had a premonition that something major was going to happen that morning. But he probably didn't realize just how major the event would turn out to be.

After several hours of surveying the area for fossils, the two men had agreed to search one last gully before heading back to camp. Johanson spotted a tiny piece of bone that he suspected was the arm from a fossil hominid skeleton. He bent down to pick it up.

It was hominid, all right. No sooner had he picked it up than he noticed yet another piece of hominid bone nearby—and another and another.

Both Gray and Johanson realized immediately that they had stumbled onto something big. Most hominid fossil skeletons consist of a couple of scraps of bone, perhaps as much as a full skull in a few exceptional cases. But it rapidly became apparent that this latest find was a remarkably complete hominid skeleton; there were bones and fragments of bones all over the place. And all of them appeared to be from the same specimen.

The two men jumped into their Land-Rover and raced back to camp in a frenzy of excitement. "We've got it!" Gray yelled to a passing group of scientists. "Oh, Jesus, we've got it. We've got the whole thing!"[6]

By afternoon, an army of people from the camp—anthropologists, archaeologists, and other workers—had gathered in the gully to dig out the pieces of the skeleton. They worked through the night. Rock music blared from a small tape recorder; the Beatles' song "Lucy in the Sky with Diamonds" played over and over again.

By morning, the fossil had acquired a nickname: *Lucy*, after the Beatles' song. She was the most complete skeleton ever found of any

Lucy, who lived perhaps three million years ago beside a lake in what is now Ethiopia, is the most complete specimen of an early hominid yet found.

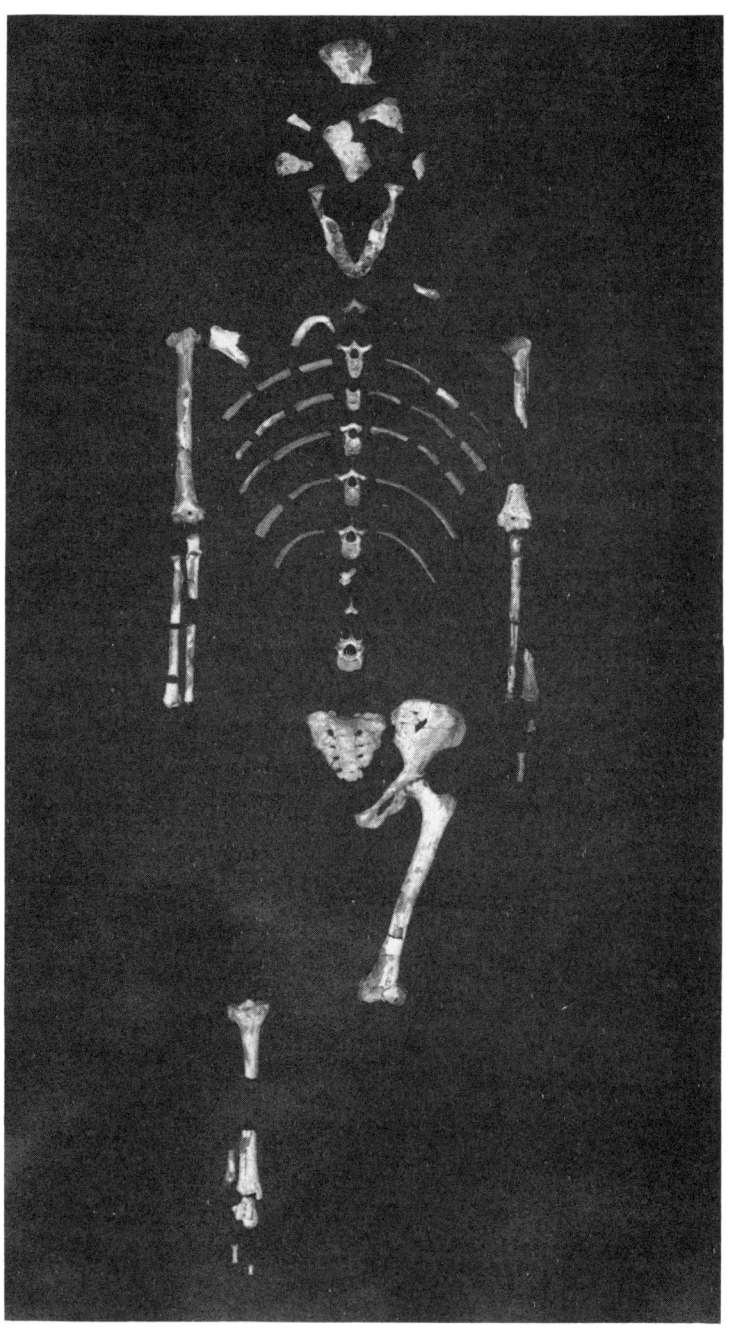

hominid species other than *Homo sapiens*. Seventy percent of her bones had survived the ravages of time.

But what was Lucy? Was she genus *Homo*? *Australopithecus*? How old was the skeleton?

The answers to those questions would rock the paleoanthropology world over the next four years. And Lucy was by no means the last major find at the Hadar dig.

The next year Johanson and several others, including a French film crew and a *National Geographic* photographer, found an amazing collection of hominid fossils on a hillside not far from where Lucy had been found. They were the bones of many different hominids. In fact, they were initially thought to represent a single family of hominids that had been killed in some long ago disaster; they were consequently given the nickname the "First Family." More likely, however, they were a group of unrelated fossil specimens that had simply accumulated in one spot, perhaps dragged there after being eaten by predators.

Were the First Family of the same species as Lucy? They didn't seem to be at first. They were larger and less delicately constructed. Both Lucy and the First Family were small by modern human standards, but Lucy was the smallest of them all, standing barely 3 feet (0.9m) tall.

When Mary and Richard Leakey saw Johanson's fossils, they offered the opinion that the First Family were early members of the genus *Homo*, in keeping with the family belief that *Homo* was a very old genus indeed. Lucy, so obviously different in size from the members of the First Family, was probably a member of a different species, they felt, possibly *Australopithecus*. As it happened, Mary Leakey had recently discovered several fossil hominids at her own site in Laetoli, Tanzania, that appeared to be of the same species as the First Family. It was her belief that those fossils were of the genus *Homo,* too.

At the time, Johanson agreed, but he was soon to change his mind. His colleague, Tim White, who had worked for several years with Mary Leakey at Laetoli, convinced him that Lucy and the First Family were in fact all members of the same species. Why did they look so different? Because the First Family fossils were all males, and Lucy was a female.

Male and female animals of a single species often look quite different, especially in size. This phenomenon is called *sexual dimorphism.* The size difference between Lucy and the First Family was unusually large, but White and Johanson decided that the resemblances between the fossils were more important than the differences.

But what species were they? Johanson and White decided that they were not of the genus *Homo* after all, as the Leakeys believed, but *Australopithecus.* Furthermore, they were not members of any previously known type of *Australopithecus* but members of a brand-new species, which they called *Australopithecus afarensis,* after the Afar Triangle region where the Hadar site was located.

The Leakeys were not thrilled when they learned that Johanson and White had decided that the First Family were australopithecine. But what really angered the Leakeys, and Mary Leakey in particular, was that Johanson and White had also decided that the fossils Mary had found at Laetoli, which she believed to be of the genus *Homo,* were also specimens of *Australopithecus afarensis.* To make matters worse, Johanson announced this decision at a scientific meeting shortly before Mary Leakey was to announce her own discovery.

Where does *Australopithecus afarensis* fit on the hominid family tree? When the Hadar fossils were dated, they turned out to be the oldest hominid fossils yet discovered—possibly as old as 3.5 million years.

In fact, it is the belief of Johanson and White that *Australopithecus afarensis* is the common ancestor of *all* later hominids, including both the later australopithecines and the genus *Homo.* Lucy may well be the oldest of all hominids; surely she is the oldest yet discovered. The family tree in Figure 7 illustrates the Johanson view of the hominids.

The Hadar find yielded an amazing collection of fossils: ranks of chimpanzee skulls (rear); remains of the First Family, Lucy and other Australopithecus afarensis *fossils, including a knee joint. At bottom left are bone fragments from Laetoli.*

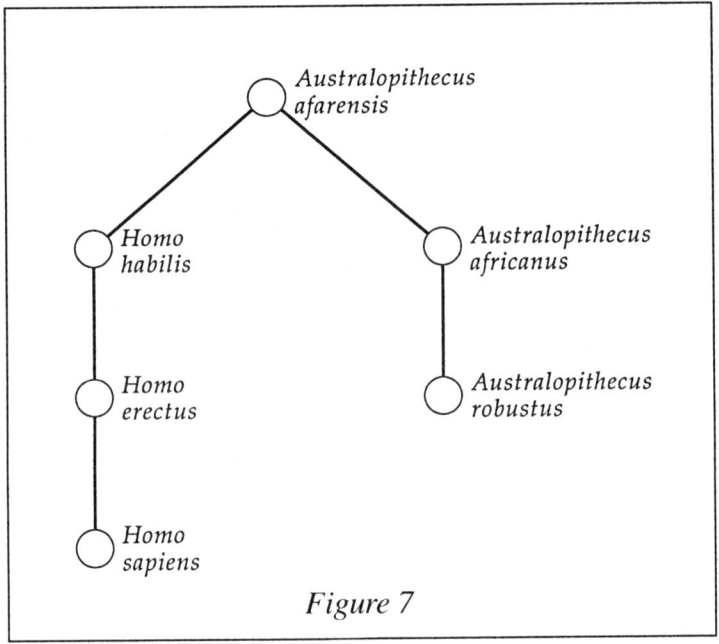

Figure 7

Obviously, this family tree contradicts the Louis Leakey hominid family tree, as described in the last chapter. But neither does it follow the single-species-hypothesis. It represents a compromise between the two views, with the later australopithecines becoming collateral relatives—cousins—of the genus *Homo*, but with everyone descending from a single australopithecine ancestor.

This family tree is controversial. The Leakeys, in particular, do not agree that it is accurate. However, at the present time, it may be the best one available. Nonetheless, it will almost certainly be changed as new, more dramatic fossil finds are made and the details of human evolutionary history become even clearer.

ONE DETAIL that remains surprisingly obscure in the drawing of the hominid family tree, even after many significant fossil finds, is the evolution of our own species, *Homo sapiens sapiens*. When did modern human beings arise? Did we descend directly from *Homo erectus?* Or was there an intermediate ancestor, such as *Homo sapiens neanderthalensis?*

This is a subject on which paleoanthropologists have changed their minds several times in the last century. Originally, *Homo sapiens neanderthalensis* was thought to be a cousin of modern humans rather than a direct ancestor. Then the opinion arose that perhaps *Homo sapiens sapiens* did descend from the neanderthals. Now the pendulum has swung back again, and the neanderthals are considered to be cousins, a group that became extinct some time in the last hundred thousand years.

Neanderthals, incidentally, were not the stumbling, stupid brutes that they were once pictured as being. Paleoanthropologists are fond of saying that if you gave a Neanderthal man a shave and a bath and dressed him up in a coat and tie, he could pass as a typical twentieth-century businessman.

The erroneous belief that the neanderthal was primitive in form grew out of two historical accidents. Several early fossils of neanderthal represented specimens that had been crippled with arthritis, and an early attempt to reconstruct a neanderthal skeleton was done poorly, leading to an inaccurate picture of this early modern human. Surprisingly, neanderthals actually had larger brains than modern humans, though this is probably because their bodies were larger overall, not because they were more intelligent than *Homo sapiens sapiens* (though this last point may be open to debate). We'll have more to say about this in the next chapter.

If modern humans did not descend from the neanderthals, where did we come from? Fossils of essentially modern human beings have been found in Africa dating from more than 100,000 years ago, though evidence of large numbers of *Homo sapiens sapiens* (as opposed to neanderthals) date back only about 50,000 years. However, there are fossils of *Homo sapiens* from as long ago as 400,000 years that show distinctly modern features, such as rounded skulls indicating large brains.

The major controversy in paleoanthropology concerning *Homo sapiens sapiens* is whether all modern human beings descended from a single *Homo erectus* ancestor or whether modern human beings evolved several times from *Homo erectus.* These are known as the replacement theory and local continuity theory, respectively.

The fossil record is ambiguous on this point, but this may be one area where the molecular anthropologists that we met in Chapter Five can come to the rescue. No matter how modern humans evolved from *Homo erectus*, all human beings on earth today must have had a common ancestor at some point in history. This is why, as pointed out in the first chapter of this book, we are all related to one another.

The real question is, how long ago did this common ancestor live? If the common ancestor was recent, say, between 100,000 and 200,000 years ago, then it would seem that all modern human beings must have descended from a single *Homo erectus* ancestor, as the replacement theory suggests. But if the common ancestor existed in the far distant past—if, for instance, it is one of the earliest members of the species *Homo erectus*, from roughly 1.5 million years ago—then modern human beings must have arisen several times from *Homo erectus* stock, as the local continuity theory suggests. And that's why it is a job for the molecular anthropologists, who so neatly solved the riddle of the common ancestor of hominids and apes by comparing the genes of humans, chimpanzees, and gorillas.

To pinpoint the common ancestor of modern humans, a group of researchers has taken samples of DNA from 147 people around the world, analyzed that DNA, and put the results into a computer, which constructed a family tree that pinpointed the last common ancestor shared by those individuals.

The type of DNA used in this study was not the DNA from the chromosomes found in the nuclei of the cells. Rather, it was a special type of DNA called *mitochondrial DNA*. The mitochondria are organs (or organelles) found within cells that produce the energy that drives the processes that keep the cells alive. Some years ago, it was noticed by observant biologists that the mitochondria resemble tiny bacteria, and it is now believed that they were once separate organisms that joined together in a mutual relationship with the cell billions of years ago. The mitochondria supply energy for the cell, and the cell gives the mitochondria a safe place to live; thus, the relationship is to the benefit of both.

One of the clues that mitochondria were once separate organisms is that they have their own DNA, separate from the DNA of the cell. And the mitochondria reproduce separately from the cell itself, splitting in two, like bacteria, passing copies of their DNA to both copies of themselves. When a cell divides, half of the mitochondria enter one copy and half enter the other.

Just as the DNA in the nucleus of the cell slowly mutates with time, so the mitochondrial DNA mutates; for this reason, it can also be used as a molecular clock. In fact, it is believed to be a more accurate molecular clock than the nuclear DNA.

Human beings reproduce sexually. When a baby is conceived he or she receives nuclear DNA from both the mother and father; this DNA is mixed together in a way that gives every human being genes that are slightly different from the

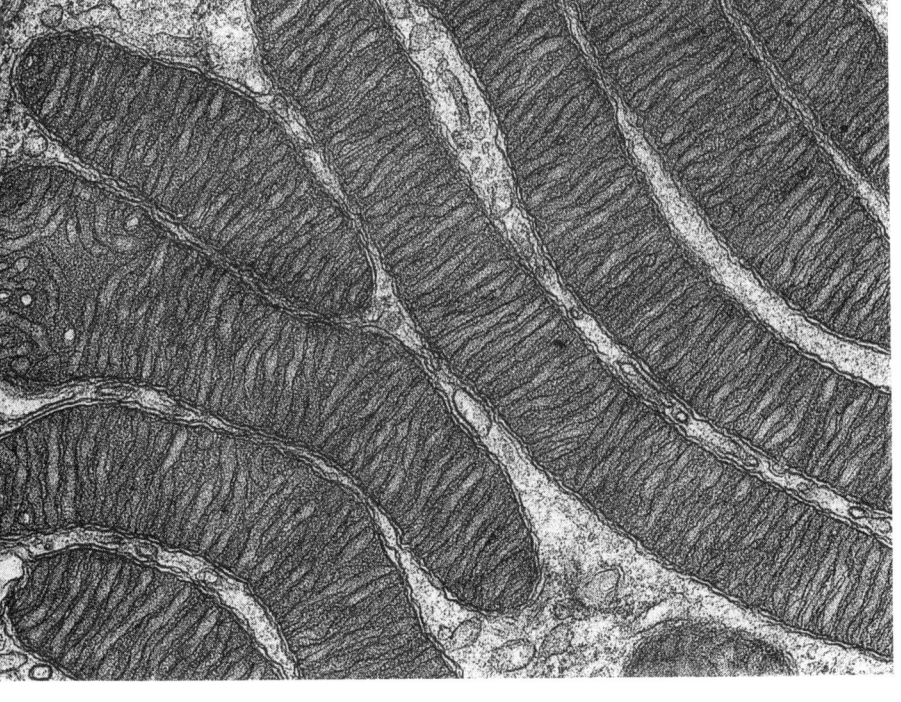

It is now believed mitochondria (pictured here) joined in a mutual relationship with the cell billions of years ago.

genes of every other human being. This makes the process of comparing nuclear DNA between individuals highly confusing. After all, how can you tell which differences in the genes, since the last common ancestor shared by these individuals, are due to sexual reproduction and which are due to mutational changes? But mitochondria do not reproduce sexually; in theory, each mitochrondrion (sing. of *mitochondria*) should have identical DNA. In practice, this means that all differences are due to mutations, and thus the mutations can be used as a molecular clock.

Human beings, like most sexually reproducing animals, receive their mitochrondria only from the mother; the sperm cell of the father is too small to hold these tiny organelles. Thus, when we study changes in mitochondrial DNA in order to pinpoint a common ancestor, we can only pinpoint *female* ancestors.

What did this study find out? *All* of the individuals in the study had descended from a common ancestor, just as expected, and almost certainly that common ancestor was a woman who had lived in Africa between 285,000 and 143,000 years ago. The molecular anthropologists like to refer to this woman as *Eve,* after the biblical mother of all humans. Is there an Adam to go with this Eve? Surely there must have been, but the mitochondrial DNA can tell us nothing about him, because as we said, mitochondria are only passed on by mothers, not fathers.

This would seem to support the replacement theory, which says that modern humans descended from a single *Homo erectus* ancestor. However, this is not really so. It is possible that Eve's immediate descendants were all members of *Homo erectus* and were not themselves modern humans but somehow became separated from the rest of the *erectus* population. Then several of Eve's descendants may have given birth to separate modern populations of *Homo sapiens sapiens,* as the local continuity theory suggests. Perhaps further study will resolve this ambiguity. Still, this glimpse of our evolutionary past, based purely on the information held within the cells of our living bodies, is immensely exciting.

PART FOUR

THE WAY WE ARE

EIGHT

UPRIGHT AND QUITE BRIGHT

What are the two features that set human beings apart most dramatically from the rest of the animal world?

If you've been paying close attention in earlier chapters, you should have no trouble with this question. These features are our upright posture—the way we stand and walk with our heads directly above our hips—and our large brains. No other mammal is like us in either regard.

This is not to say that our closest relatives, the African apes, can't walk erect when they want to, or that they are unusually stupid. Nothing of the sort.

In fact, gorillas and chimpanzees are capable of short bursts of erect walking, or running. They just don't do it habitually, like we do. And they don't do it particularly well. They aren't built for it.

The African apes are also strikingly intelligent, considering that they aren't human. In fact, they are quite probably the most intelligent members of the animal kingdom, *Homo sapiens* excepted.

When we look back at the history of hominids, we can see that erect posture and large brains have been consistent themes in our evolutionary history. All hominids walk erect. And all members of the genus *Homo* have large brains, though *Homo sapiens* has a larger average brain size than *Homo erectus*, and *Homo erectus* has a larger average brain size than *Homo habilis*.

But paleoanthropologists are not content simply to note this fact. They also ask why. Why have hominids developed these two unusual features? What evolutionary advantage did we gain by standing up straight and being smart?

The answers may seem obvious, particularly in the case of intelligence. If we weren't smart, we wouldn't be able to build houses, cars, telephones, and all the other things that protect us from the elements, help us get where we are going, and help us communicate with one another. But these are recent developments. We have to bear in mind that natural selection did not know that we would eventually create the advanced technology we have today. Evolution cannot plan ahead. Why did *Homo habilis*, who lacked houses, cars, and telephones, have a bigger brain than *Australopithecus afarensis*, who may have been *habilis's* direct ancestor? And why did *afarensis* walk upright?

Early paleoanthropologists believed that intelligence came first and then erect posture. Their reasoning was that intelligent creatures would have made tools and that standing erect would have left their hands free to carry those tools. For instance, a hunter carrying a bow and arrow would need to walk erect and hold the bow and arrow in two hands in order to shoot it. A chimpanzee would have considerable trouble trying to run and shoot a bow and arrow at the same time.

However, it eventually became apparent that erect posture had preceded large brains by at least a million, and perhaps even several million, years. Lucy, who lived 3.5 million years ago, stood perfectly erect, but she had a brain only slightly larger than that of a chimpanzee. There is no evidence that Lucy had anything approaching humanlike intelligence, or that she used tools of any kind.

Ironically, Lucy and the other australopithecines may have been even better at walking erect than modern humans are. The evidence for this comes from a remarkable series of fossil hominid footprints discovered by Mary Leakey at Laetoli, Tanzania.

These footprints appear to have been left by australopithecines, probably members of *Australopithecus afarensis,* walking through fresh volcanic ash that subsequently became wet and hardened. The footprints are beautifully preserved, almost as though they had been made a few weeks ago rather than 3.6 million years ago, and careful study of these footprints, as well as examination of fossil skeletons, has told paleoanthropologists a good deal about how early hominids walked.

A trail of hominid footprints uncovered at Laetoli, Tanzania, demonstrates that human ancestors walked upright 3.75 million years ago.

The results seem to suggest that early hominids had a pelvic structure better adapted to erect walking than our own. Why? Ironically, because they had smaller brains. The larger heads of human babies require that mothers have an unusually large space between their hips for the baby to emerge at birth. This enlargement has actually been detrimental to erect posture, which may explain the human tendency to develop backaches.

Why did erect posture evolve in the first place? This is something of an evolutionary mystery, and there is, as yet, no definitive answer.

It has been suggested that erect posture may have been a reaction to changing climatic conditions. The early primates lived in the trees, at a time when forests were plentiful. But when the forests began to disappear and be replaced by grasslands, many primates (including our own prehuman ancestors) became adapted for life out of the trees. It has been suggested, therefore, that walking erect allowed early hominids to see for long distances while walking in tall grass. However, it is now believed that erect walking developed before our ancestors moved from the protection of the forests to the open grasslands, where they would have been vulnerable to a large number of predators.

Perhaps these early hominids needed their hands free to carry something other than tools. One possibility is that they used their hands to carry food. Human beings are unusually social animals—that is, we live and interact in groups made up of other members of our species, and we share the work of food gathering among us. Perhaps hunters needed their hands free to carry food home from the hunt. Or, those who gathered vegetables needed to carry armfuls of roots and other edibles.

Alternatively, mothers may have needed their arms free to carry their babies. This is less of a problem for other primates, because the babies can cling to the mother's fur. But humans have very little in the way of fur, and human mothers must cling tightly to the child to prevent it from falling to the ground. However, the loss of fur is believed to be a relatively recent development. Lucy was probably covered with fur, as most probably were early members of the genus *Homo*, though this fur may have been shorter than that of other primates. Since fur does not fossilize, though, this is difficult to determine. And this leads to the general question of why we lost our fur in the first place, a question that has never been definitively answered.

TAKING FOR GRANTED the fact that our ancestors had become fully erect walkers by at least 3.6 million years ago, why then did advanced intelligence begin to evolve a million or so years later, with the advent of *Homo habilis*?

Unlike erect posture, intelligence leaves no direct mark on the fossil record; brains are not preserved. Nevertheless, there are many indirect signs of intelligence available to the paleoanthropologist.

One of these is the size of the braincase, the area in the skull where the brain is housed. Roughly speaking, the larger the braincase, the larger the brain.

Does greater brain size always mean greater intelligence? Not necessarily. There are several species of animal on earth today with larger brains than human beings, such as elephants and whales. Are these animals more intelligent than we are?

No. What really matters is the ratio of brain size to body size. A large portion of brain mass is devoted to controlling the body; thus, the larger the body, the larger the brain, regardless of intelligence. Elephants and whales have large brains because they have large bodies; people have large brains because we're unusually smart.

One way to measure relative brain size is the *encephalization quotient,* or *EQ*. Basically, this can be looked at as the size of the brain divided by the size of the body, though in practice, the method by which the EQ is determined is a lot more complicated. The EQ is a pretty good gauge, overall, of a species' intelligence (though it tells us nothing about the intelligence of individual members of the species, even if their brains are smaller or larger than the species' average). And humans have the largest EQ of any animal.

Brain size is used as the dividing line between *Australopithecines* and the genus *Homo*. The range of brain sizes found among modern humans is rather wide, roughly from 700 or 800 cubic centimeters (cc) to almost 2,000 cc. Thus, any hominid that has a brain size within this range is considered to be *Homo; habilis* just makes the cutoff. (Actually, Louis Leakey had to fudge this definition slightly to fit his *habilis* specimen into the genus.)

Homo habilis brain sizes begin at just under 700 cc. *Homo erectus* brain sizes begin at about 800 cc and expand to more than 1,100 cc in Peking man, one of the very last specimens of *erectus*. Thus, brain size has expanded over the last two million years, though the jump in brain size from *erectus* to modern humans is still rather great, with an average brain size of about 1,400 cc in *Homo sapiens sapiens*. (As we mentioned in the last chapter, the average brain size of *Homo sapiens neanderthalensis* is actually slightly larger than the average brain size of *homo sapiens sapiens*, but the EQ is almost exactly the same.) It should be noted, however, that brain size tells us little about the internal structure of the brain, which is also a powerful determinant of intelligence.

Another fossil sign of intelligence is tools. Humans are not the only tool users on earth. Hungry chimps, for instance, use wooden sticks to lure juicy termites out of their nests. But we are the only ones who use sophisticated techniques to make a variety of tools. This tradition seems to have begun with *Homo habilis;* there is no sign of any earlier hominid making stone tools.

Paleoanthropologists like to talk in terms of *tool kits* and *tool industries*. Tool kits are collections of tools often found together at certain fossil sites, including tools for cutting, chipping, smashing, etc. Tool industries are patterns of tools in kits that seem to have persisted over surprisingly long periods of time. There are indications that each industry is associated with, or at least originated with, a different species of the genus *Homo*.

For instance, the *Oldowan industry* (named after Olduvai Gorge, where some of the best samples of Oldowan tool kits have been found) seems to have been originally associated with *Homo habilis*, though it survived that species' extinction and didn't disappear until about 200,000 years ago. The tools in the Oldowan kit have been given names by paleoanthropologists, such as the hammerstone, the unifacial chopper, the bifacial chopper, the flake scraper, and so forth. These tools were mostly created by striking stones together so as to cause the stone to break apart into flakes. The flakes were used as knifelike scrapers; what remained of the stone could be used as a hammer or weapon.

The *Acheulian industry*, which began almost a million years ago, seems to have originated either with *Homo erectus* or the very earliest *Homo sapiens*. It existed alongside the Oldowan technology for half a million years and vanished about 150,000 years ago.

The Acheulian technology was similar to, but in some ways more advanced than, the Oldowan. Tools in the Acheulian kit have been given names such as the cleaver, the pick, and the spheroid. But the real star of the industry was the hand ax,

Acheulian stone axes (left) and a collection of Mousterian stone tools (right) are significant indirect indicators of early humans' intelligence.

which came in several forms. The hand ax was an extremely versatile tool that would have been well suited for slicing through tough animal hide. The Acheulian tools themselves sometimes seem more primitive in appearance than those of the Oldowan, but the methods that were used to create them were apparently more advanced, much less wasteful of raw material.

Finally, the Acheulian industry was replaced about 150,000 years ago by the *Mousterian culture*, which is associated with *Homo sapiens*. Some of the Mousterian tools are so beautifully made that they may have been used primarily in formal rituals. And this, in turn, would seem to indicate a new plateau in human intelligence, an ability to think beyond tangible needs to the needs of the spirit.

All of the tools that have been found at prehistoric sites have been made out of stone; hence, the term *Stone Age* for this lengthy period of human evolution. However, this does not mean that all tools during these periods were made out of stone; rather, this was the only material that survived into modern times. Probably, there were also tools made from wood and perhaps even from animal bone; many tools may have been constructed from a combination of materials, with stone used only for the most sturdy parts and the cutting edges. And, in fact, some indirect evidence of bone and wooden artifacts have been uncovered alongside early *Homo sapiens* remains.

At one time paleoanthropologists believed that tools were one of the driving forces behind the evolution of intelligence. Early humans developed tools, then became more intelligent in order to make better use of those tools. As they became more intelligent, they made even better tools, which required them to become even more intelligent to use them better, etc. Now, however, a new driving force is suspected behind the development of intelligence—the need of human beings to better understand other human beings.

As we said earlier in this chapter, humans are unusually social animals. We live in social groups with other members of our species, divide up responsibility for various tasks among members of the group, and educate our offspring to a greater degree than other animals.

These social interactions are an important part of what it means to be human, and early humans who were particularly skilled in social interactions may have been more evolutionarily successful than those who were less skilled. Among other things, being skilled in social interactions meant being able to communicate effectively (hence, the evolution of speech), being useful to the group (hence, the development of specialized skills), and so forth. Most important, being skilled in social interactions meant having some understanding of one's fellow human beings.

Perhaps the greatest of all unsolved mysteries in the evolution of human intelligence is the mystery of consciousness or self-awareness. The remarkable thing about human intelligence is not just that we can think subtle and complex

The Museum of Anthropology in Nairobi, Kenya, displays a re-creation of a typical group of men and women who inhabited that part of Africa about two million years ago.

thoughts—though this is certainly remarkable—but that we are *aware* that we are thinking subtle and complex thoughts. We are not only smart, we are *conscious* of our smartness.

This may seem obvious to you, but why should it be so? It is not hard to imagine a very advanced computer programmed to "think" as well as a human does, even to respond emotionally as human beings do. Yet, would such a computer be aware of its own thoughts? Would it be aware of its own emotions? Would it be conscious?

The answer, of course, is that we do not know. We cannot enter the mind of a computer—indeed, we do not even know if a computer can have a "mind" to enter, in the sense that humans have minds—and so we cannot know if it would be aware or not. But we tend to assume that it would not be. A computer may be able to think, but it seems unlikely that it would be aware of its own thoughts.

In a sense, human beings are computers programmed by evolution. Our brains are extremely sophisticated machines, capable of doing things that no other machine on earth is yet capable of doing.

Why, then, don't we feel like machines? Why aren't we just automatons, rapidly executing complex programs that tell us how to survive in a complex world? What evolutionary value is there in consciousness, in self-awareness?

No one knows. In fact, the question may not have an answer, at least not one that we would recognize if we found it. Science cannot even tell us what consciousness is, much less why it exists. Perhaps someday we will understand it better, but for the present, we are almost entirely in the dark.

Are other animals conscious? There is no way at present to know, though it is tempting to believe that other mammals are, at least. Reptiles, birds, and fish may also be conscious, though they are farther away from us in evolutionary terms. It seems unlikely that insects, plants, and bacteria are conscious, though we cannot prove that they are not. Certainly viruses, which are not even alive in the conventional sense, are not conscious. Or are they?

Probably consciousness is an intrinsic part of the nervous system, so only those organisms with some kind of brain, however rudimentary, are likely to be conscious. Some scientists have suggested that consciousness is a primate development, resulting from the need of our tree-dwelling ancestors to coordinate complex sensory data in the brain. Consciousness, then, may be a method for maintaining order among a massive influx of data.

It may even be that consciousness is a purely human trait. One reason that consciousness has evolved may be the same reason that intelligence evolved, as a way of helping humans to deal with one another. If we were not conscious, if we were not aware of our own existence, then it would be impossible for us to imagine what it would be like to be someone other than ourselves. We could not empathize with other human beings. We could not work as well together as we do to make society function.

Perhaps consciousness, more than intelligence, is the greatest gift that we have received from evolution. If so, which species do we have to thank for this gift? One of the human family? *Homo habilis? Homo erectus?* Or our own species, *Homo sapiens?*

Whenever consciousness arose, and however many species we share it with, it is ultimately this combination of consciousness and intelligence that allows us to look back across the gulfs of time that separate us from our most distant ancestors and ask the most fundamental of all questions: Where did we come from? How did we come into existence? How did we come to be . . . human?

NOTES AND SOURCES USED

1. Quoted in Roger Lewin, *Bones of Contention* (New York: Simon & Schuster, 1987), 110.
2. Quoted in Lewin, *Bones of Contention*, 105.
3. Richard Dawkins, *The Blind Watchmaker* (New York: Norton, 1986), 273.
4. Quoted in Donald Johanson and Maitland Edey, *Lucy: The Beginnings of Humankind* (New York: Simon & Schuster, 1981), 90.
5. Quoted in Lewin, *Bones of Contention*, 138.
6. Quoted in Johanson, *Lucy: The Beginnings of Humankind*, 17.

Asimov, Isaac. *Asimov's Biographical Encyclopedia of Science and Technology.* New York: Avon, 1972.

Brace, C. Loring, and Ashley Montagu. *Human Evolution: An Introduction to Biological Anthropology.* New York: Macmillan, 1977.

Calvin, William H. *The River that Flows Uphill.* New York: Macmillan, 1986.

Campbell, Bernard. *Human Evolution.* Chicago: Aldine, 1974.

Cann, Rebecca L. "In Search of Eve." *The Sciences* 27 (September/October 1987): 30–37.

Dawkins, Richard. *The Blind Watchmaker.* New York: Norton, 1986.
Eldredge, Niles, and Ian Tattersall. *The Myths of Human Evolution.* New York: Columbia University Press, 1982.
Gowlett, John. *Ascent to Civilization: The Archaeology of Early Man.* New York: Knopf, 1984.
Gribbin, John, and Jeremy Cherfas. *The Monkey Puzzle.* New York: Pantheon, 1982.
Johanson, Donald, and Maitland Edey. *Lucy: The Beginnings of Humankind.* New York: Simon & Schuster, 1981.
Leakey, Richard. *The Making of Mankind.* New York: Dutton, 1981.
_____and Roger Lewin. *Origins.* New York: Dutton, 1977.
_____.*People of the Lake: Mankind & Its Beginnings.* New York: Doubleday, 1978.
Lewin, Roger. *Bones of Contention.* New York: Simon & Schuster, 1987.
_____.*Human Evolution: An Illustrated Introduction.* New York: W. H. Freeman, 1984.
_____."Modern Human Origins Under Close Scrutiny." *Science* 239 (11 March 1988): 1240–1241.
_____."Molecular Clocks Turn a Quarter Century." *Science* 239 (5 February 1988): 561–563.
Pilbeam, David. "The Descent of Hominoids and Hominids." *Scientific American* 250 (March 1984): 84–96.
Reader, John. *Missing Links.* Boston: Little, Brown, 1981.
Stringer, C. B., and P. Andrews. "Genetic and Fossil Evidence of the Origin of Modern Humans." *Science* 239 (11 March 1988): 1263–1268.

RECOMMENDED READING

There are many popular books available on the subject of human evolution, some of them written by the very paleoanthropologists who developed the theories that they are writing about. Alas, the information in these books dates quickly, and books published more than ten or fifteen years ago are often hopelessly behind the times. The following is a list of books that the author of the present volume has found entertaining or useful. However, only volumes from 1976 on have been included, to keep the list reasonably current. Be warned that some of this material has already become dated, though not quite as severely as in earlier books.

The Selfish Gene by Richard Dawkins (Oxford University Press, 1976).
The Blind Watchmaker by Richard Dawkins (Norton, 1986).

Dawkins is a British zoology professor who writes entertainingly and knowledgeably on the subject of evolution. Though not specifically on human evolution, these two volumes provide a briskly written introduction to the basic concepts of natural selection, especially as they relate to genetics. *The Selfish Gene* is about the role of the gene in evolution; *The Blind Watchmaker* is about the ways in which natural selection can produce complex organisms that could never arise by random chance. Both are well worth reading.

The Monkey Puzzle by John Gribbin and Jeremy Cherfas (Pantheon, 1982).

Gribbin, a physicist and an astronomer, and Cherfas, a biologist, may seem an odd pair to write a book on paleoanthropology. However, their subject is the fusion of hard science and social science represented by the new molecular anthropology, and their particular combination of expertise is appropriate to the story they tell. The volume was written during the final years of the controversy between the traditional paleoanthropologists and the "upstart" molecular anthropologists. Gribbin and Cherfas offer an impassioned argument in support of the molecular clock discovered by Vincent Sarich and Alan Wilson. This book is a good source of material for readers interested in this subject.

Lucy: The Beginnings of Humankind by Donald Johanson and Maitland Edey (Simon & Schuster, 1981).

Ostensibly the story of Johanson's discovery of *Australopithecus afarensis* (a.k.a. Lucy) and the First Family at Hadar, but actually a delightful inside look at what the life of a highly successful paleoanthropologist is really like. Fun to read and immensely informative. Includes a brisk and intelligently written capsule history of the field from Darwin's time onward.

Origins by Richard Leakey and Roger Lewin (Dutton, 1977).
People of the Lake: Mankind & Its Beginnings by Richard Leakey and Roger Lewin (Doubleday, 1978).
The Making of Mankind by Richard Leakey (Dutton, 1981).

Richard Leakey, son of Louis and Mary Leakey, has become not only the premier paleoanthropologist of our time but also the premier explainer of paleoanthropology. These three books, written by Leakey (sometimes in collaboration with his friend, noted science writer Roger Lewin) over the course of only four years, all cover pretty much the same territory, but all are worth reading. *People of the Lake* is the most readable, *Origins* the most detailed and best illustrated, and *The Making of Mankind* (based on a television series with which Leakey was involved) the most up-to-date. Be warned, however, that the information on *Ramapithecus* and the early history of the hominids in these books is already dated. See Roger Lewin's book *Bones of Contention,* below, for more current information.

Bones of Contention by Roger Lewin (Simon & Schuster, 1987).

A history of the major controversies of twentieth-century paleoanthropology, which is pretty much the same thing as a history of twentieth-century paleoanthropology. Lewin covers in considerable detail the debate over Dart's identification of *Australopithecus,* the battle between the supporters of *Ramapithecus* and the molecular anthropologists, the various controversies involving Louis and Richard Leakey, the KBS tuff controversy, and so forth. The emphasis here is more on people and politics than on human evolution itself; an excellent followup to some of the other books in this list.

Human Evolution: An Illustrated Introduction by Roger Lewin (Freeman, 1984).

A short and solid, if rather dry, summary of current knowledge in paleoanthropology by one of the most prolific and distinguished authors in the field. Well organized and illustrated in the style of *Scientific American* magazine (which is from the same publisher). Low on anecdotal material but heavy on information; for the reader who wants to learn the basic facts without wading through half a dozen books to find them.

Missing Links by John Reader (Little, Brown, 1981).

Probably the best popular history of paleoanthropology available. Reader supplies a wealth of detail and intelligent comment on almost every major discovery yet made in the field, from *Homo sapiens neanderthalensis* and *Australopithecus afarensis* to the footprints at Laetoli. If the briefer histories supplied by Johanson and Leakey (above) whet your appetite for more of the same, look for it here.

INDEX

Page numbers in *italics* refer to illustrations.

Acheulian industry, 137, *138*, 139
Adenine, 40
Africa, 70
African apes, 53, 71, 72, 81, 83, 130
Amphibians, 66
Animal (*animalae*) kingdom, 28
Animals, 22
Antibodies, 81
Apes, 52–53, 64, 68, 71
    African, 53, 71, 72, 81, 83, 130
    Asian, 71, 73, 81, 88
    common ancestor with man, 45, 48, 49, 52, 64, 71, 74, 76, 85–88, 123
    East Indian, 53

Apes (*continued*)
    posture of, 22, 57, 76, 129, 131
    spinal cord, 55–57
    See also specific names
Arboreal primates, 68, 70
Asian apes, 71, 73, 81, 88
Atoms, 24, 38, 82
Australopithecus, 86, 96, 98, 100, 112, 117, 119, 131
    brain size, 135
Australopithecus afarensis, 117, *118*, 119, 130, 132
Australopithecus africanus, 55, 56, 57–58, 103, 104, 111
Australopithecus boisei, 101, 102, 104, 106
Australopithecus robustus, 58, 103, 104

Babies, human, 133, 134
Backaches, 133
Bacterium, 22–24, 40, 41, 42

Bats, 68
Bible, 45
Binomial nomenclature, 27, 28
Biochemists, 82, 86
Biology, 31
Black, Davidson, 59
Bone tools, 139
Brain, 20, 54, 60, 70, 98, 131, 134
    of Mesozoic mammals, 67
    of modern man, 70, 129, 142
    size, and intelligence, 135–136
Braincase, 55, 135
Broom, Robert, 57–68

Carbon-14 dating, 109
Carnivores, 102
Cells, 24, 42
Cenozoic era, 66
Chimpanzees, 64, 65, 71, 76, 129, 131
    use of tools, 136
China, 53, 58
Choppers, 137
Chordata, 28
Chromosomes, 38, 39, 80, 84, 89
    point mutations in, 84–85
Classes, 27, 28
Classification of organisms, 24–29, 47–48
Common ancestor:
    of hominids, 119
    of man and ape (missing link), 45, 48, 49, 52, 64, 71, 74, 76, 85–88, 123
    of modern men (*homo* genus), 122–126

Communication, 140
Computers, 88–90
Consciousness, 140–144
Continents, 70–71
Cousins, 15–16, 18, 20, 22
Creation, 30, 45
Cro-Magnon man, 52
Cytosine, 40

Dart, Raymond, 55–58
Darwin, Charles, 31, 33–36, 34, 44, 48, 53, 60
Dating of fossils, 108–110
Dawkins, Richard, 90
Dawson, Charles, 54
Dinosaurs, 67
DNA (deoxyribonucleic acid), 38–42, 80, 84, 123
    mitochondrial, 123–126
Dolphins, 68
Dubois, Eugene, 52–53, 54

Earth, 68
    age of, 30–31, 48, 75
    evolving, 70
Ecological niche, 67, 68, 102
Emotions, 20
Environment, 67
Eoanthropus dawsoni, 54
EQ (encephalization quotient), 135
Eras, 66
Erect posture, 20, 22, 63–64, 76, 129, 130–134
"Eve" (man's common ancestor), 126
Evolution, 32, 35, 43, 45, 66–71
Extinction, 48, 91
Eyesight, 68, 70

Families, 27, 28, 47
Family tree, 16, 18, 19, 20, 29

Family tree (*continued*)
    nodes, 30, 43, 47
"First Family"(fossils), 116, 117
Fish, 66
Flake scraper, 137
Food gathering, 134
Footprint fossils, 131, 132
Foramen magnum, 55, 57
Fossils, 10, 48, 49, 52, 53, 54, 55, 57, 60, 86, 87, 106, 131, 132
    dating of, 108–110
    skeletons, 49, 98, 114, 131
Fur, 76, 134

Genealogy. *See* Family tree
Generic names, 26, 27
Genes, 36–42, 81, 82, 87
    of apes and man compared, 82–84
Genetic code, 81, 82
Genetics, 31, 36–42, 81
Genus, 26, 27, 47, 57
Geologists, 30, 110
Gibbons, 71
God, 30, 45
Gorillas, 64, 65, 71, 72, 76, 83, 129
Gray, Tom, 113, 114
Guanine, 40

Hadar region, Ethiopia, 113
Hammerstone, 137
Hand ax, 139–141
Hands, 70, 75, 76
Hierarchical system, 29–30
Hominidae family, 47, 57, 63
Hominids, 63, 64, 71, 72, 75, 88, 91, 102, 106, 114, 130
    erect posture of, 63–64, 130–134

Hominids (*continued*)
    family tree of, 103–105, 112, 119, 120
Hominoidea super family, 64, 77
Homo erectus, 53, 58, 63, 101, 103, 104, 120
    brain size, 130, 132, 136
    homo sapiens descent from, 120, 126
    tools, 137
Homo genus, 47, 63, 96, 98, 100, 101, 103, 111, 116, 117, 119
Homo habilis, 101, 102, 103, 104, 106, 111, 130, 132
    brain size, 136
    tools, 136, 137
Homo sapiens, 43, 47, 104, 122
    defined, 28
    descent from *homo erectus*, 120–126
    extinct species of, 48
    subspecies of, 63
    tools of, 139
    See also Homo sapiens sapiens; Man, modern
Homo sapiens neanderthalensis, 49–52, 63, 120–121, 138
Homo sapiens sapiens, 52, 101, 120–122, 126, 138
Hrdlicka, Alex, 74
Humans. *See* Man, modern
Hunting, 134

Insectivores, 68
Intelligence, 130–131, 134, 138, 139, 144
    and brain size, 135

Java man, 53. *See also Homo erectus*

Jaws, 54, 74, 75
Johanson, Donald, 113–119

KBS tuff, 108, 110, 111, 112
Kingdoms, 28
Knuckle walking, 76

Laetoli, Tanzania, 117, 118, 119, 131
Lake Turkana, Kenya, 106
Language, 76, 140
Leakey, Jonathan, 100, 111
Leakey, Louis, 95–101, *97*, 104, 108, 116, 117
Leakey, Meave, 106, *107*
Leakey, Mary, 95, 98, 99, 100, 116–117, 119, 131
Leakey, Richard, 106, *107*, 108, 110, 111
Lewis, G. Edward, 71–74
Life, origin of, 42
Linnaeus, Carolus, 24–27, 30, 45, 47, 48, 71
Local continuity theory, 122–123
Lucy (fossil), 113–119, *115*, *118*, 131

Mammals, 28, 66, 67, 68
Man, modern, 20, 22
  common ancestry with apes, 45, 48, 49, 52, 64, 71, 74, 76, 85–88, 123
  common ancestry of men, 122–126
  posture of, 20, 22, 129, 130, 133
Mendel, Gregor, 36, *37*
Mesozoic era, 66–68
"Missing link" (common ancestor of man and ape), 45, 48, 49, 52, 64, 71, 74, 76, 85–88, 123

Mitochondria, 123–126
Modern Synthesis, 38
Molecular anthropologists, 66, 91, 122, 123, 126
Molecular biology, 31, 82
Molecular clock, 84–87, 91, 124
Molecular dating, 82–91
Molecules, 10, 24, 38, 40, 82
  protein, of man and ape compared, 82, *83*, 84, 85
  self-replicating, 66
Monkeys, 22, 68, 71, 81
Mousterian culture, 139, *140*
Museum of Anthropology, Nairobi, *143*
Mutation, 33, 36, 81, 124
  point, 84–85

Names and naming, 19. See also Classification of organisms; Nomenclature
Natural selection, 33–36, 38, 42, 43, 45
Neander Valley, Germany, 49
Neanderthal man, 49–50, *50*, *51*, 52, 121
Nervous system, 143
Neutral mutations, 84
Nocturnal animals, 67, 68
Nodes, 16, 18, 20, 30, 47, 48
Nomenclature, 26–29. See also Classification of organisms
Nuttall, Henry Falkiner, 81, 82

Oldowan industry, 137, 139
Olduvai Gorge, Tanzania, 95, *99*, 100, 102, 111, 137

]156[

On the Origin of Species by Natural Selection, or the Preservation of Favoured Races in a Struggle for Life (Darwin), 33, 44, 48
Orangutans, 64, *65*, 71, *73*, 88
Orders, 27, 28
Organisms:
    earliest living, 42
    multicelled, 42–43
    single-celled, 66

Paleoanthropologists, 10, 54, 64, 77, 81, 85, 86, 98, 101, 104, 108, 111, 139
Paleontologists, 10, 31, 64
Paleozoic era, 66
Paranthropus, 58
Peking, China, 58, 59
Peking man (Sinanthropus pekinesis), 58–59
    brain size, 136
Pelvic structure, 133
Phyla, 27, 28
"The Phyletic Position of *Ramapithecus*" (Simons), 74
Physical characteristics, 36
Pilbeam, David, 75
Piltdown man, 54
Pithecanthropus erectus, 53. See also *Homo erectus*
Plant (*plantae*) kingdom, 28
Point mutations, 84–85
Polynomial nomenclature, 26, 27
Posture, erect, 20, 22, 60, 63–64, 129–134
    pelvic structure and, 133
    reasons for evolution of, 133

Potassium/argon dating, 109–110
Primates, 68
    ancestral, 68–71
    arboreal, 68, 70
    See also Apes
Prosimians, 70
Protein molecules, 82, *83*, 84, 85

Radioactive dating, 108–110
Ramapithecus, 71–77, 85–86, 87–88
Relatives, 15–22
Replacement theory, 122, 126
Reptiles, 66, 67–68
Rituals, 139
Rock of Gibraltar, 49

Sarich, Vincent, 82–88
Scraper, 137
Self-awareness, 140–144
Sexual dimorphism, 117
Sexual reproduction, 33, 124
Shrew, tree, *69*
Simons, Elwyn, 74, 75, 86
Sinanthropus pekinensis (Peking man), 58–59
Single-species hypothesis, 103, 119
Sivapithecus, 87–88
Siwalik Hills, India, 71
Skeleton, fossil, 49, 98, 114, 131
Skull 1470, 105–108, *107*, 110–112
Social interaction, 133, 140
Solo River, Java, 53
Somme River, France, 49
South Africa, 55, 57
South America, 70

Species, 22
    common ancestors between, 30, 35, 45
    extinct, 48, 91
    nomenclature, 27
    subspecies, 27
    variations in, 33–35
Speech, 76, 140
Spinal cord, 55
Stone Age, 139
Stone tools, 49, 98, 100, 137, *140*
Strata, earth, 108, 110
Surnames, 19

Taungbaby. See Australopithecus africanus
Taxonomy, 24
Technology, 76, 130
Teeth, 75, 100, 102
Terminal nodes, 18, 43, 47
Thymine, 40
Tool industries, 136, 137
Tool kits, 136–138

Tools, 60, 70, 98, 100, 131, 137, 138
    Acheulian, 137–139
    bone, 139
    Mousterian, 139, *140*
    stone, 49, 98, 100, 137, *140*
Tree-dwelling primates, 68, 70

Variation, within species, 33–35
Vegetarianism, 102
Vision, 68, 70
Volcanic rock, 110

Walking erect. *See* Posture, erect
Whales, 68
White, Tim, 117, 119
Wilson, Allan, 82–88
Wooden artifacts, 139

Zinjanthropus boisei, 100, 101